REEDS
WEATHER
HANDBOOK

More titles in the Reeds Handbook series

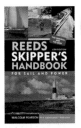

*Reeds Skipper's Handbook
for Sail and Power*
6th edition
Malcolm Pearson
978-1-4081-5629-2

Reeds Crew Handbook
Bill Johnson
978-1-4081-5571-4

Reeds Knot Handbook
Jim Whippy
978-1-4081-3945-5

REEDS WEATHER HANDBOOK

FRANK SINGLETON

SHERIDAN HOUSE

Published by Sheridan House
4501 Forbes Boulevard, Suite 200, Lanham, Maryland 20706
www.rowman.com

10 Thornbury Road, Plymouth PL6 7PP, United Kingdom

Distributed by NATIONAL BOOK NETWORK

Copyright © 2014 by Frank Singleton
This American edition of *Reeds Weather Handbook*, First Edition,
is published by International Marine by arrangement with
Bloomsbury Publishing Plc.

British Library Cataloguing in Publication Information Available

Library of Congress Cataloging-in-Publication Data Available

ISBN 978-1-57409-354-4 (cloth)

**For more weather information visit Frank Singleton's website at
http://weather.mailasail.com/Franks-Weather**

Dedication

*To my family and all other sailors from whom
I have learned so much over so many years –
and still continue to learn*

Acknowledgements

Photographs are by the author unless stated otherwise. Satellite images are reproduced courtesy of the Dundee Satellite Receiving Station, University of Dundee, UK.

Thanks are due to Reeds' editorial staff for their help in making a difficult topic readable. Also to my wife for her apparently inexhaustible patience during the writing.

CONTENTS

Sources of weather information 99

Getting forecasts 113

Observing – learning by experience 120

Appendix 126

Index 142

Introduction

Weather is both fascinating and frustrating – there are few absolutes, and little is ever really certain. Words such as 'always' or 'never' can rarely be used, and there are few hard and fast rules that are of any use to the sailor. Complaints about forecasts are usually the result of the weather being so unpredictable on all scales of time and space. Nevertheless, marine weather forecasts are continually improving and, used sensibly, make a significant contribution to safety at sea.

This book focuses on the practicalities of obtaining, understanding and using advice from weather forecasts, personal observation and experience. Racing sailors, whether round the cans or round the world, use forecasts to improve their chances of winning; but for most of those going to sea, the primary objective of using weather information is to minimise risk.

As we all know, forecasts are not – and can never be – precise or exact. Using professionally produced forecasts to minimise risk when sailing is best achieved through the addition of experience and common sense. This requires a pragmatic and realistic appreciation of the strengths and weaknesses of weather prediction.

The necessity of using marine forecasts issued for safety purposes is paramount. Equally, this book emphasises the value of being able to use other weather information to help you understand and utilise the 'official' forecast services. In meteorology, as in sailing, you never stop learning.

Meteorology is a complex subject and any attempt to delve into the theory in a pocket-sized book would lead to sweeping and misleading generalisations or partial truths. That is why theory has been kept to the minimum necessary to give a basic understanding of the background to forecasts and their production.

Good luck and fair weather. If not fair weather, then I hope that your foul weather is well forecast!

Author's note

Abbreviations and acronyms abound in the meteorological and marine worlds. All those used, and a few more, are listed in the Appendix, as are definitions of some scientific and technical words.

Although there are frequent references to my former employer, the UK Met Office, most statements refer to national weather services in other countries.

URLs of websites are not given in the text because they are all too prone to change. A good search engine should find the sites mentioned.

Understanding air masses

On any one day when sailing, more often than not it will be in the same kind of air mass throughout. Changes of air mass occurring at fronts can be dramatic and the cause of bad weather; we need to know about them. Many sailors will try not to be at sea when fronts are expected, so encountering them will be comparatively rare. For much of the time, it will be more useful to be aware of the kind of air mass to be expected when sailing, and the weather most likely to be experienced.

Textbooks sometimes go into great detail with different categories and sub-categories but, for our purposes, we can take a rather more simple approach, as in this diagram.

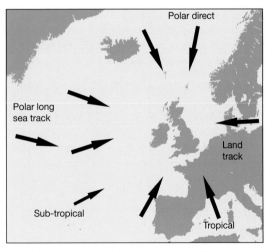

Fig 1.1 *The main types of air mass around the UK*

Air directly from polar regions

Stating the obvious, this air will be cold. Perhaps less obviously, being cold it cannot hold as much water vapour as can air from warmer areas of the world, even though it must pass over relatively warmer seas to reach wherever we are sailing. The **tropopause** will be comparatively low and this limits convection from the sea surface to heights of 8 or 9km, while tropical convection may reach heights of 15 or 16km. Further, this air is not much affected by man-made pollution. These facts have several consequences.

◆ Convection from the relatively warm sea surface will lead to **showers**, although these will not be as heavy as in air from other directions.

◆ With the low water content and relatively weak convection, thunderstorms will be infrequent.

◆ Any man-made pollution will be well mixed by the convection and the visibility will be generally excellent – except in showers.

◆ Strong gusts are likely and the 'weight of the wind' effect may be marked.

◆ Wind strengths will generally be near to the speed implied by the pressure gradient.

◆ Particularly in winter, groups of showers can merge in areas of low pressure known as polar lows.

The 'weight of the wind' may be a somewhat illusory effect; but like much in meteorology, there is some truth in the idea. Cold, dry air is denser than warm, moist air. For a given wind speed, there can be 10 per cent more force on the sail of a yacht in cold air than in warm air. However, in cold air there will be gusts in the wind.

An increase of wind speed from 15 to 17 knots will create an increase in force on the sail of about 28 per cent; an increase from 15 to 20 knots will give an increase in force of nearly 80 per cent. This gust effect may make weight of wind more of an apparent (rather than a real) contribution to the force of wind on sail.

Air indirectly from polar regions

When there is a large, slow-moving low pressure area in high latitudes, air from the pole can have a long sea track before it reaches waters off the west coasts of Europe or North America. This creates some significant differences compared to air coming directly from polar regions.

First, the air is considerably modified having been warmed by the sea, and can contain more water gained by evaporation from the sea surface. A second effect is that of large lows, that is, areas where the air is generally rising. There can be troughs of low pressure where the rising effect is enhanced. The air will not have picked up much man-made pollution but will have more salt from sea spray and breaking waves. The sailor will see several differences compared to air coming direct from the pole:

◆ There will be more and heavier showers.

◆ Thunderstorms will be more likely.

◆ Visibility will be good, except in showers, but the air will be less clear.

◆ Strong gusts will be likely with a greater chance of squall lines.

In cold air, clouds will mainly be due to convection, like these (see Fig 1.2).

Fig 1.2 *Convective cloud in polar air*

Air from the subtropics

Air originating from the subtropical high pressure areas will reach western Europe or the west coast of North America after a long sea track. High pressure areas, such as the Azores high and its Pacific counterpart, are where the air is generally descending, and are on the polewards side and part of the Trade wind circulation.

Because it has been descending, this air is dry and warm through much of its depth due to **adiabatic** effects. However, in the lowest 1,000m or so, it is both cooled and made more humid by contact with and evaporation from the sea. Thus, above the moist, relatively cool low-level air, the temperature often increases. This is known as a temperature inversion.

There can be no convection of any consequence from the surface – because the sea is cooler than the air immediately above it. The temperature inversion acts as a lid, restricting any turbulent mixing of air to the air

Tropical air

space below that level. Salt particles from the sea will be trapped below the inversion and, as the air is moist, it is likely that small water droplets will form. Together with the lack of convective mixing, this type of air mass will be characterised by the following:

◆ Poor visibility, with the possibility of mist or fog.

◆ Steady winds, with speeds considerably lower than might be expected from the pressure gradient.

◆ The possibility of low cloud thick enough to give steady, mainly light precipitation with uniformly small drops. This is known as **drizzle**.

The rather featureless conditions with low-lying cloud might look like this.

Fig 1.3 *Low cloud in warm, moist air* © *David Holbourn*

Tropical air

Particularly significant around western Europe is the warm, moist air that can be found over Spain and France. This air will have been heated by the ground sufficiently to cause vigorous convection due to high surface temperatures, leading to heavy showers and thunderstorms. These can be the reason for the breakdown of hot dry spells in late summer or autumn. Over North America, the effect is most marked when such air coming from the south meets a cold front coming from a westerly point, leading to extremely vigorous convection and tornadoes.

Air mass with a land track

The effects of air coming over land are too variable to go into in great detail here in a way that would help sailors. In coastal waters to the west of mainland Europe and North America, air coming from the east will often be fairly dry but heavily polluted. Much will depend on whether this air has come round a high-pressure area to the north, or from a low-pressure area to the south. In the first case, the air will have been descending and there will be a low-level temperature inversion trapping the pollution near to the surface. In the second case, it is more likely to be moist enough to give **rain** or showers.

The three most important points are:

◆ Fog will not occur in a polar air mass, except occasionally in the Arctic.

◆ Showers will not form in an air mass coming from the subtropics unless it has passed over heated land.

◆ Air from low latitudes passing over land heated by the sun can yield very heavy showers.

Fog

Fog and mist are both formed by airborne water droplets; the distinction is only in the level to which visibility has been reduced. For marine (and aviation) purposes, fog is defined as less than 1,000m visibility and mist is visibility from 1,000m to 2nm. Over land, for highway warning purposes, fog is regarded as 200m or less.

The word 'haze' is used when the visibility is reduced by solid particles, e.g. smoke and industrial pollutants. The distinction is because mist may deteriorate into fog but haze should not, unless there is a change in the source of the air.

Fog formation causes some confusion because many of us are accustomed to the fog that occurs over the land on clear nights with little wind. The ground cools by radiation and cools the air just above it down to its **dewpoint**, when water starts to condense onto the ground as dew or in the air as small water droplets. In effect, this creates a cloud at ground level and is only likely to be of interest to sailors when it drifts down valleys and estuaries. In such cases, the sea will usually be warm enough to disperse the fog. This is called 'radiation fog' and it may well be 'burned' off by the sun.

Fog at sea is usually 'advection fog', and occurs when air from an area of relatively warm sea moves over a colder area. The air will cool by contact with the sea and, if it reaches the dewpoint temperature, then condensation will take place. This can happen at any time of the day throughout the year and with a wind blowing, as long as it is not too strong. In fact, wind helps to maintain the fog as it will continue to bring warm, moist air over cooler waters. Sea fog is not readily burned off by the sun. With strong winds, sea fog may lift into low cloud but visibility at sea level will remain poor.

Moist air can be lifted over the coast to give hill fog. Although not a hazard for the sailor, it indicates that the air temperature is close to dewpoint at sea level and that sea fog may be possible.

Sea fog areas

Certain areas are particularly prone to sea fog and some of the better known ones are listed here:

◆ Around western and south-western approaches to the British Isles and the Atlantic coast of France, with air coming from the direction of the Azores high.

◆ Atlantic coasts of Spain and Portugal with south-west winds, especially in the summer. The so-called Portuguese Trade Winds (northerlies) maintain cold water down much of this coast. The fog can fill the Rías Baixas; once formed, the cold water in the Rías helps the fog to persist for days or even weeks at a time.

◆ Eastern coastal waters of the UK. This is due to water over the middle of the North Sea always being a degree or so warmer than near the coast of Britain. North or north-east winds onto the north of East Anglia and east or south-east winds onto coasts from the Wash up to the Moray Firth can all be misty with fog. Sometimes this lifts into low cloud and is known as haar or sea fret.

◆ Near Danish and German North Sea coasts, in the winter and spring when these waters are very cold, so that a westerly wind is very likely to give sea fog.

◆ Western coasts of the USA; San Francisco and Los Angeles fogs are well known.

Fog

◆ The Grand Banks and eastern coastal waters of North America generally have a strong gradient of sea temperature. Any air coming from the south round to the east is a candidate for fog.

◆ Central and eastern parts of the south coast of Arabia in the south-west monsoon, known as the 100-day fog. The strong winds cause overturning of the sea and upwelling of cold water, while the winds bring air from a warm sea.

◆ Near Europa Point, Gibraltar, when the east-going tide brings in cold water and the wind is bringing warmer Mediterranean air from the east. Particularly in the winter and spring, this may extend along the Spanish coast.

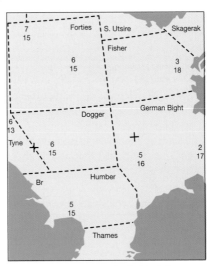

Fig 1.4 *Sea temperatures in the North Sea*

A knowledge of average sea temperatures can help to indicate areas where sea fog is likely. For areas around the British Isles, the RYA/Royal Met Society 'Met Map' indicates where, when and with what wind directions sea fog is possible. This map is available online, free, from the RYA. The values shown are average sea temperatures in September and February, the months with the warmest and coldest times of the year. They illustrate why sea fog occurs in south-west winds up the English Channel and around eastern coasts of Britain.

The charts here show the average frequencies of sea fog around the British Isles and North Sea coasts. Comparing these with the Met Map temperatures provides a demonstration of the reasons discussed above.

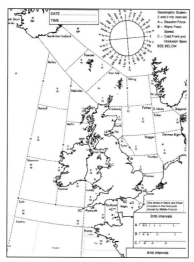

Fig 1.5 *An RYA Met map*

Fog

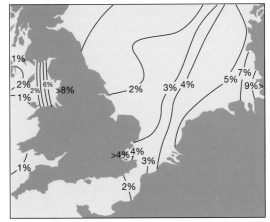

January

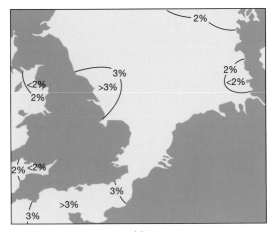

July

Fig 1.6 *Sea fog frequency*

Other causes of sea fog

Fog over the sea can occur for other reasons:

◆ Rain on a front can give enough rise in humidity to cause fog to form.

◆ Thunderstorms, particularly those that drift up from France in the late summer, can decay but leave areas of moist air with a high dewpoint. Fog can occur with little or no warning when these patches drift over cold water, such as near the Channel Isles.

◆ Arctic sea smoke, when there is convection into very cold air from water just above freezing temperatures.

Fog prediction may not be good and there is little a slow-moving vessel can do if caught out at sea. Careful use of radar and an AIS Class B transponder will give reassurance and be a great help, but won't guarantee safety. Conventional advice is to get into shallow water, if possible, to avoid large ships. On the other hand, that might increase the risk of collision with smaller vessels that do not carry or use their AIS. Also, the chance of encountering wooden vessels that do not show up on radar is increased. At least if you are in or near the shipping lanes then you should see the dangers on your AIS, and be seen because of your own AIS signal.

Showers and thunderstorms

Convective (cumulus) clouds form when air warmed by the surface rises like bubbles from the bottom of a heated saucepan. As the air bubbles rise, they expand and cool adiabatically. If they cool down to the dewpoint temperature, water will start to condense to form water droplets that we see as cloud. The height at which the temperature of the rising air reaches the dewpoint temperature is known as the **condensation level**. As is seen in Figs 1.7 and 1.8, the condensation level is virtually constant over a wide area.

Condensation releases **latent heat** that gives the clouds more buoyancy and they continue to rise as long as the cloudy air is warmer (or less dense) than the surrounding air. Small cumulus clouds, as in Fig 1.7, do not usually give any showers.

Just occasionally, small cumulus clouds give very light showers. This is happening in Fig 1.8 and is more likely late in the day when convection is weakening. Small water droplets may then have enough time in the cloud to grow large enough to fall out of the cloud bottom. More usually, rising bubbles reach the cloud top and evaporate as they emerge.

In the same way that we cannot predict just where and when the first bubbles will rise from the bottom of the pan of heated water, it is impossible to say where or when the next cumulus cloud will form. With vigorous convection, these clouds get deeper and the risk of significant showery activity, perhaps with thunderstorms, increases. However, this can only happen when the cloud tops freeze. Careful observation will show whether that is happening or not.

Fig 1.7 *Small cumulus early in the day*

Fig 1.8 *Shallow cumulus can give light showers*

Fig 1.9 *These fairly large cumulus did not give any rain*

In Fig 1.9, although the clouds were quite deep and temperatures at their tops were well below 0°C, no freezing had taken place. Cloud droplets can often remain liquid with temperatures as low as -10°C to -20°C and occasionally as low as -40°C.

Bubbles pushing out of the tops can be seen to be fragmenting as the droplets evaporate in the clear air round the cloud. The fragmenting bubbles subside down the side of the cloud. That can be seen just left of centre.

The first sign of cloud tops freezing is when the emerging bubbles stop evaporating. That can be seen when they no longer fragment and subside but start to spread out and the cloud tops take on a slightly fuzzy appearance.

In Fig 1.10, the freezing process is just starting to happen. This has several consequences:

◆ More latent heat is released as the drops freeze, giving a boost that will take the tops higher up, often to the tropopause.

Fig 1.10 *Cloud tops starting to freeze*

◆ The ice crystals evaporate far more slowly than water drops and spread out to form an anvil shape under the tropopause temperature inversion.

◆ Ice crystals grow faster than water drops in the cloud and some quickly get big enough to fall rapidly, collecting water drops on the way down to give heavy rain or hail showers.

◆ The freezing of the water drops can cause splintering, creating many small ice particles.

◆ These processes lead to separation of electric charge so that the cloud top becomes positively charged and the bottom negatively charged.

◆ In turn, that induces a positive charge at sea or ground level.

◆ If the charge separation is large enough, there will be a risk of lightning strikes.

AIR MASSES

Lightning

Fig 1.11 shows how the electric charges are distributed in a thunder cloud usually known as **cumulonimbus**. In Fig 1.12, the top of the cloud is mostly composed of ice crystals and a heavy shower is falling. There would almost certainly have been thunder and lightning.

Lightning protection

Lightning is a spark between positive and negative charges, rather like holding wires from two battery terminals near each other, but on a gigantic scale. Contrary to popular belief, there is no difference between sheet and forked lightning. When seen through cloud, the light is diffused to give 'sheet lightning'; when seen directly, it is 'forked lightning'.

The chances of being hit by lightning are fairly low, but it is a serious hazard for any vessel at sea. When lightning strikes a boat, the greatest effect is often the loss of all electronic equipment, including hand-held devices, even when not in use. A sensible precaution at the first sign of trouble is to put cellphones, hand-held GPS, VHF, laptop PCs, tablets,

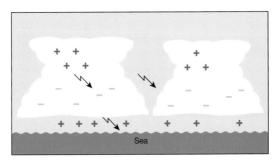

Fig 1.11 *Charge separation*

Fig 1.12 *A mature thunderstorm cloud © Maggie White*

personal EPIRBs, etc in the oven. This acts as a Faraday cage and should protect those items.

Obviously it is best to avoid being at sea when thunderstorms are expected, but that is not always feasible. It used to be recommended that chains were carried to protect vessels in a thunderstorm, the idea being to wrap the chains around the mast and trail

Fig 1.13 *Lightning © Roy Conchie*

Fig 1.14 *A lightning master might save money*

them in the sea to ensure that any strike would quickly be dissipated. That is not an option for a small boat.

A more practical approach is to use a passive device, as pictured in Fig 1.14, resembling a wire bottle brush and bolted or riveted to give a good contact to the mast-head. There needs to be a good connection from the mast or its rigging to a good, solid earth, say, via a sacrificial anode or grounding plate. The theory is that this prevents a build-up of charge on a single point by diffusing the charge over the many points and allowing it to leak away to earth.

On the basis of one very close encounter, the author thinks they may well work – sometimes, at least…. At a low cost, it is a solution worth serious consideration.

Avoiding the risk

Lightning can be both frightening and dangerous, but it is sometimes possible to minimise the risk. One option is to stay in harbour whenever heavy showers are in the forecast, particularly when the forecast mentions thunder. However, even that is not 100 per cent safe;

even with other taller masts around, you may just be the unlucky one.

In any case, that would often keep you in port unnecessarily. In a shower cloud, and whenever there is rain involving freezing of water drops, there will always be some charge separation. This might or might not be sufficient to give lightning.

Some guidance is available in output from weather prediction models, known as the CAPE parameter. This is a measure of instability and is available using several GRIB services (see the Appendix). CAPE is an indicator of how much energy can be released in a storm, assuming there is enough heating to trigger convective clouds in the first place. Values are expressed as joules per kg of air but all we need think about are the numerical values.

When the CAPE index is zero, the air will be stable and convection is not possible. When CAPE is below 1,000, risk of heavy showers is real but small. When CAPE exceeds 1,000 there will be an increasing chance of heavy showers, as the table shows.

CAPE index	How unstable
< 1000	Slightly
1,000–2,500	Moderate
2,500–3,500	Very
3,500+	Extremely

Pictorial displays, like the example overleaf (Fig 1.15), can be more intuitive. In the blue areas, thunderstorms are most unlikely. A significant risk starts in the areas coloured deep yellow and increases as the red colours

Other dangers of thunderstorms

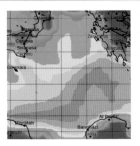 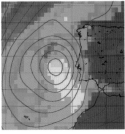

Fig 1.15a and b *CAPE – a lightning risk indicator*

deepen. In Fig 1.15a, the prudent sailor would try to avoid the area east of Tripoli, near the coast of North Africa. Synoptic charts showed a rather weak-looking trough of low pressure. In Fig 1.15b a low to the west of the Iberian Peninsula has a rather diffuse area of moderate activity.

> A visual warning can sometimes be seen in the grandly named altocumulus castellanus. These small, turreted, medium-level clouds can be the leftovers of a thunderstorm but show that the air might still be sufficiently unstable to give more storms.

Other dangers of thunderstorms

In a heavy shower, the rain or hail drags air downwards at great speed so that it spreads away from the cloud in all directions. The result can be strong gusts of 30–40 knots or more. In extreme cases, there may be a line of thunderstorms moving with the winds at about 3km above ground level. The result can be a line of squalls. First signs might be a roll of cloud in a distinctive line.

Fig 1.16 *Risk of more thunderstorms to come*

Fig 1.17 *A squall approaching © Maggie White*

Other dangers of thunderstorms

Fig 1.18 *Line squall getting close* © Duncan Galloway

When convection is particularly vigorous, there can be the serious hazard of a waterspout. The most serious of these are associated with tornados but in Europe these are extremely rare. Most likely conditions are when there is vigorous convection just ahead of a cold front moving south-eastwards across the USA. Like tornadoes over land, these are severe hazards by any standard. Waters such as the Florida Keys and the Great Lakes are particularly prone.

Less serious are the so-called fair-weather waterspouts. These occur with vigorous thunderstorms in, apparently, preferred areas. Waters around the Isle of Wight seem particularly prone to waterspouts, although they have been seen elsewhere in British waters – near the Isle of Man, for example. The warmer waters of the Mediterranean must always make waterspouts a possibility, but it is difficult to select any particular area as more prone than elsewhere.

Fig 1.19 *Waterspouts are dangerous* © *Matt Rooney*

These fair-weather waterspouts may not be as dangerous as their American tornado counterparts but can still be dangerous and avoiding action should be taken. On the radar, they will show up as a hard, distinctive echo. Fig 1.19 shows a waterspout seen in the Aeolian Islands.

> As ever in meteorology, little is ever simple or straightforward. Thunderstorms and lightning can also occur close to active weather fronts, especially cold fronts or occlusions. Listen to the forecast; at any mention of heavy or thundery rain, take a look at CAPE. You may not always get it right but you will lessen the chances of getting it wrong.

Frontal lows and their formation

The large-scale winds around high and low pressure areas determine the kind of air mass you will experience when sailing. As these lows and highs move, the direction of the air flow will change and the air mass will change. Sometimes this can be a gradual process lasting a day or more; at other times there can be sudden changes at fronts associated with lows that generally move from west to east.

Many books and websites describe how fronts are formed, how they behave and their relationship to the **jet stream**. Inevitably, these descriptions are highly idealised and simplistic; no two lows, highs or fronts will be identical and there can be marked differences between them.

Here, we will concentrate on what the sailor sees and experiences without much discussion of the theory. Some general idea of fronts is assumed. Typical textbook diagrams of the birth and development of a frontal low look like this:

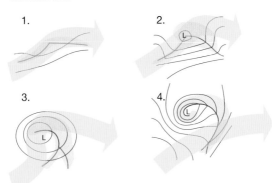

Fig 2.1 *The life cycle of a frontal wave*

Frontal lows and their formation

To clarify the terminology used by meteorologists, the terms 'low', 'low pressure area' and 'depression' are synonymous in this context. Similarly for 'high', 'high pressure area' and 'anticyclone'. The word 'cyclone' is usually reserved for tropical storms.

At **1**, a wave starts to form on a front in association with a slight bend in the jet stream. As the low moves eastwards, it will change its appearance as shown at **2**. The wave and the jet stream both become increasingly distorted. The region between cold and warm fronts is known as the warm sector. At **3**, the cold front is catching up with the warm front and the occlusion process has started. By this time, the jet stream has become quite convoluted.

At **4**, the occlusion process has been completed and the jet stream has re-formed south of the low centre. To the north of the jet stream, the high-level winds will have formed a vortex over the surface low pressure area.

Fig 2.2 *Change is coming!*

Frontal lows and their formation

Many lows crossing the North Atlantic and North Pacific form on the polar front, the boundary between air from the polar areas and air from the subtropics. Frontal lows that reach western Europe often start off at the eastern seaboard of North America.

Much of the energy in a frontal low comes from the supply of latent heat released as the warm, moist air in the warm sector is lifted and cooled. After the occlusion process, the low will be cut off from this source of energy. Left to itself, the low will then slowly fill up, largely due to friction at the surface, although it may be followed and absorbed by further lows.

To help your understanding of frontal lows:

◆ Low pressure areas are where the air is generally rising, but far more slowly than in convective clouds.

◆ Most of the rising air is concentrated around the fronts.

◆ The rising leads to cooling, creating cloud and rain.

◆ The high-level jet stream winds will blow high-level cloud away from and ahead of the warm front.

◆ The jet stream will be positioned more or less along the cold front and blow high-level cloud along the cold front.

The last two points explain why high cloud gathering from the west or north-west is usually seen ahead of a warm front, while there is often a sharp clearance after a cold front.

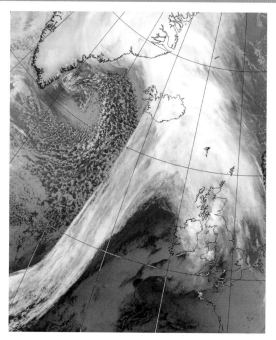

Fig 2.3 *Behind cold fronts, clouds often clear sharply; ahead of warm ones, cloud increases steadily*

This is illustrated in the satellite image in Fig 2.3 of a well-marked cold front in mid Atlantic. The much more amorphous-looking shield of high cloud on the warm front is streaming across the north of the UK and the North Sea. The lower-level cloud of the warm front is hidden by the high cloud. This infrared image shows the convective cloud behind the cold front and the absence of any such cloud in the warm sector. It shows well the random nature of the convection.

What you may see at sea level

Having seen a view from space, let us now see it from sea level.

From the general synopsis in a shipping (offshore) forecast or from a synoptic chart, a sailor should know that a low pressure area and its fronts are approaching. Watching the sky may give early confirmation. Before any high-level **cirrus** clouds start to form, the air at high levels will be rising and becoming more humid. That may result in aircraft condensation trails forming and persisting in the moist air, rather than evaporating quickly as they will often do when the air at those levels is dry.

A more reliable sign will be wispy-looking high cirrus cloud spreading from the north-west. This cloud will be mostly comprised of ice particles. If this cloud spreads quickly, it shows that the jet stream winds are strong and that the approaching system has a lot of

Fig 2.4 *Condensation trails might presage a warm front, but…*

energy and could become a vigorous low with strong winds at the surface. The hooks at the end of these cirrus clouds are typical.

If the front is relatively slow-moving, the air at these levels will rise comparatively slowly and steadily. The ice particles will be fairly uniform in size and shape; the cloud will spread out to give a uniform layer of cirrostratus – layered cirrus – and there may be a halo around the sun or moon.

> Haloes form when light passes through ice crystals rather than water drops, as in a rainbow. Haloes can be coloured like rainbows, but less strongly. The red colours will be on the inside of the curve, while in rainbows they are on the outside.

Fig 2.5 … *increasing cirrus almost certainly will*

What you may see at sea level

Fig 2.6 *Warm front approaching, perhaps slowly*

After the cirrus has appeared and become a layer of high cloud:

◆ The cloud will thicken, becoming progressively lower.

◆ Winds will back to south-west, south or even south-east.

◆ The wind will increase.

◆ The pressure will fall.

◆ It will start to rain.

Ahead of a warm front, the main cloud base will have lowered but there will be broken lower cloud resulting from the rain.

The speed and extent of these effects will vary greatly; there are no hard and fast rules. Listen to the forecast, watch the sky, and watch your barometer.

What you may see at sea level

Fig 2.7 *A warm front announces its presence* © Maggie White

Just ahead of the warm front:

◆ Pressure may fall a little faster.

◆ The rain is likely to become heavier.

◆ The wind may back a little further…

◆ …with increased strength.

In the absence of any other information, assume that winds will reach at least near gale force 7 to gale force 8.

As the warm front passes:

◆ The pressure will stop falling or fall much more slowly.

◆ The wind will veer, most likely to south-west or west-south-west.

◆ The wind will probably decrease.

◆ The rain will stop or at least decrease.

What you may see at sea level

Fig 2.8 *Here the fog covers coastal cliffs*

Again, there are no hard and fast rules; frontal activity will vary greatly.

If a warm front has passed, you will be in the warm sector – between the warm and cold fronts. What happens then will depend on how far you are from the low pressure centre and how warm the air is relative to the sea temperature.

Sea fog can be shallow with clear air on coastal hilltops and cliffs. At other times it may lift to form hill fog with poor visibility, not necessarily down to fog limits at sea level.

Be prepared for:

◆ Low, featureless cloud, perhaps with drizzle.

◆ Poor visibility, mist or sea fog.

◆ Near the coast, there may be hill fog.

◆ Steady, not usually very strong, winds.

What you may see at sea level

As the cold front approaches:

◆ Pressure will probably start falling again.

◆ Wind may back a little.

◆ It will start to rain, and the rain is likely to become heavy.

◆ There may be thunder and lightning.

◆ These may lead to squalls.

The arrival of a cold front may look like Fig 2.10. Some convective cloud is just starting to form. The cloud generally will be thick but some signs of a break could be appearing.

Just behind the cold front:

◆ The cloud will break.

◆ Pressure will start to rise, perhaps quickly.

◆ The wind will veer, typically to north-west.

◆ Showers will develop.

◆ Visibility will improve greatly – except in showers.

Fig 2.9 *Sea fog can be shallow © Duncan Galloway*

What you may see at sea level

Fig 2.10 *Cold front cloud © James Galvin*

Fig 2.11 shows breaks in the main cloud mass with broken low cloud after the frontal rain has ceased.

Shortly after this, skies will clear; convective cumulus clouds will appear and showers will develop as described earlier.

Fig 2.11 *Cold front cloud breaking © James Galvin*

How much the wind veers, how quickly the pressure rises, and how long the rain lasts will vary greatly. At some fronts, the wind change can be extremely sharp, perhaps 90°, maybe as much as 150°; at others it may be barely discernible.

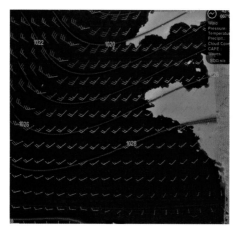

Fig 2.12 *Winds do not always follow the isobars*

The NWP output in the screenshot of a Weather4D tablet app in Fig 2.12 shows a fairly marked wind change. The actual change experienced at sea would probably have been even sharper. A forecaster using this data as a basis would have been watching the front and assessing how it was behaving. That would make it possible to give a better indication than the raw computer model shown here. The best guide to such detail as frontal wind changes is likely to be a synoptic chart produced as a combination of computer output and human interpretation.

Occluded fronts

If the cold front has caught up with the warm one, the system is said to be 'occluded'. The transition from the pre-warm front to the post-cold front conditions will take place with no warm sector. If the occlusion process has just started, the fronts will still be vigorous and the occlusion is likely to be marked by heavy rain. Thunder and lightning will always be a possibility, especially in the late summer and autumn. If the occlusion process is well advanced, the system will be weakening and the front less vigorous.

Having lost its supply of warm air, the low will deepen no further. Friction at the surface will slow the winds down. Pressure in the middle will start to rise, and the low is said to be filling. The satellite image in Fig 2.13 shows an occlusion crossing the British Isles with the parent low to the north-west. The convective activity in the cold air behind the front is clearly illustrated, as is the swirl of cloud around the low centre. By this stage, the high-level vortex will have formed and the cyclonic circulation will have extended right through the atmosphere.

As in Fig 2.3, the convective activity is clearly random. Individual clouds are a few tens of kilometres across, but there are larger groups of cloud some 100–300km in size. These larger areas might have been indicated on a synoptic chart as a trough.

Interpretation of synoptic charts

Fig 2.12 is a good example showing how much the wind can sometimes blow across the isobars near a front. More usually, as long as they are not too light, winds over the sea will more or less follow the direction of the isobars, as in Fig 2.14, though with some cross-

Fig 2.13 *Vigorous convection behind an occlusion*

isobar effect, usually moving from high to low pressure. In 1857, Buys Ballot told us that pressure will be lower to the left when standing with our back to the wind (in the northern hemisphere). This follows from the **Coriolis effect** due to the spin of the earth and is clearly shown in Fig 2.14.

This shows winds following the isobars around a ridge of high pressure and how the wind speed increases with an increase in pressure gradient.

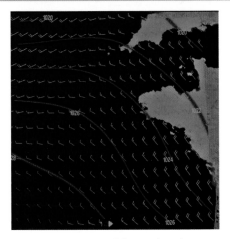

Fig 2.14 *Here, winds are following the isobars*

Surface-level winds can be estimated using the geostrophic scale shown on the UK Met Office chart. The **geostrophic wind** is the wind that would occur at a height of about 1,000m above the surface with straight isobars. At such a height it is generally assumed that there will be no effects of **friction layer**. The actual surface wind will depend on several factors, such as:

◆ Surface roughness.

◆ Curvature of the isobars.

◆ Stability of the air.

◆ The proximity of fronts.

The geostrophic scale shown here is for isobar spacing of 4 **hPa**. To use this scale, measure the spacing between two isobars. Do this at right angles to them, using navigator's dividers. Then put one point of the dividers

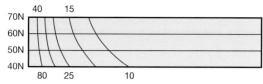

Fig 2.15 *Geostrophic wind scale, in kt for 4.0 hPs intervals*

on the left hand side of the scale at the appropriate latitude and read off the speed. The scale used should be that on the synoptic chart.

As a rough guide, expect winds to be the following fractions of the geostrophic speed:

◆ Generally, and ahead of a warm front – ⅔ to ¾.

◆ In a warm sector – ½ to ⅔.

◆ In cold air, behind a cold front or occlusion, up to full gradient speed.

Much easier is to use the GRIB file output. The wind data given is derived by taking into account the factors described above. To allow for the smoothing effect of the NWP grid spacing, add about one Beaufort force (or about 20 per cent) to the speeds.

Synoptic charts show actual and expected positions of highs, lows and fronts, thus providing an aid to understanding GMDSS forecasts. Fronts are usually in troughs of low pressure which, as we have seen, may or may not be very pronounced. Troughs of low pressure can occur for other reasons and may just be marked on a synoptic chart by a line with no symbols. Rather confusingly, these lines may just be well marked areas of cloud as seen from satellites, that the forecasters think are worthwhile showing on charts. These may or may not be associated with a trough of low pressure.

Charts fronts symbols
These are the most common symbols used for fronts. Often, these symbols will be in black; sometimes there will be other symbols, not shown here, that mark where fronts are stationary, starting to form or weakening.

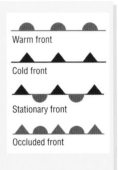

Warm front

Cold front

Stationary front

Occluded front

Non-frontal lows

In winter, there can be polar lows formed primarily by cold Arctic air moving over warmer seas, coupled with eddying around land masses such as Greenland and Iceland. These are relatively weak systems, although they may give significant falls of snow.

European sailors will be aware of the Spanish summer-heat low that deepens by day and weakens a little at night. A larger-scale effect is the monsoon-heat low over the north of the Indian subcontinent during the northern hemisphere summer.

Otherwise, the most significant non-frontal low pressure systems are tropical depressions that can develop into hurricanes, typhoons or tropical cyclones. The names used depend upon the area; meteorologically, they are basically the same. Tropical storms require warm seas – an area the size of continental USA with temperatures exceeding 26°C is sometimes quoted for North Atlantic hurricanes. These start as easterly waves and small groups of thunderstorms; in Trade wind latitudes, these storms move westwards. Some become depressions, and a few of those deepen enough to become hurricanes; as they mature, they usually turn polewards (towards the north pole in the northern hemisphere, and the south pole in the southern hemisphere).

Most leisure vessels are too slow to be able to take avoiding action when tropical storms are developing and the prudent sailor will avoid being at sea during the storm season. For the North Atlantic, this has traditionally been defined as:

> June – too soon.
> July – stand by!
> August – look out you must.
> September – remember.
> October – all over.

Like much received weather lore, the old rhyme has some truth but should not be taken too literally. In recent years there have been hurricanes in May, June, October, November and December, the longer season possibly a result of climate change. In fact, there is no month in which a tropical depression has not been seen in the North Atlantic. There is only one recorded instance of a hurricane in the South Atlantic, where the waters are colder.

Anyone sailing during the peak storm season, particularly in a tropical storm area, should keep a careful watch on weather bulletins. However, predictability is limited to a few days. If you're caught out with a tropical storm approaching, remember that they usually turn polewards. If possible, head towards the equator – assuming you have the speed to do so.

How sea breezes are formed

Most of us sail chiefly during the summer months and near the coast. Heating of the land by day and, usually to a lesser extent, cooling by night can affect both wind direction and speed. Forecasts of inshore weather sometimes mention that sea breezes are likely, but they can rarely be much more specific. Many sea breeze effects will be too small or require too detailed input to be predicted even by the best **mesoscale** NWP models. It is left to the sailor to use experience and understanding to assess and be prepared for the results.

Wind is caused by pressure differences – or gradients – and these result from temperature differences. On the large scale, with heating right through the whole depth of the atmosphere, these are the causes of Trade winds, jet streams and frontal depressions. Conversely, cooling of the air leads to high pressure. The winter Siberian high is one example.

The sea breeze results from heating on a far smaller scale in terms of both the area and the vertical extent of the heating or cooling (Fig 3.1).

A 'typical' sea breeze day starts with a clear sky overnight, a slack pressure gradient and little or no wind. After sunrise, the land will warm up quicker than the sea. That causes the air near the ground to warm up and expand, leading to a movement of air out to sea at about 100–200m above sea level. That decreases the pressure over the land and increases the pressure over the sea. That pressure difference causes the air to flow from high to low pressure at sea level to compensate for the outflow higher up. That inflow is the sea breeze.

Conversely, the cooling ground at night causes the air to contract over the land, leading to air movement from sea to land above the surface and from land to sea at the

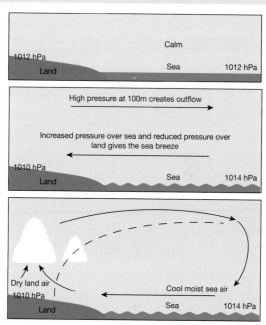

Fig 3.1 *How sea breezes form*

surface. Thus a land breeze is, in effect, a reverse of the sea breeze. In both cases, these are simplistic descriptions; the reality is greatly modified by other effects.

The sea breeze front

As the sea breeze sets in, the cool, moist air from over the sea meets warm, drier air over the land in what is sometimes called a 'sea breeze front'. The convergence of air makes the air rise, forming cloud as a result. These processes are shown in Fig 3.1. Descent of the air over the sea causes warming and makes clouds dissipate,

The sea breeze front

while the cloud at the sea breeze front moves inland as the land warms further and the sea breeze becomes stronger. This is a fairly typical example of the front moving inland, leaving no cloud over the sea.

Fig 3.2 *Sea breeze front*

Fig 3.3 *Sea breeze cloud enhanced by mountains over Cyprus*
© *James Galvin*

What affects the sea breeze?

Fig 3.3 shows a line of more than usually well-developed convective cloud at the sea breeze front. This was taken over Cyprus, with the mountains enhancing the lifting effect of the sea breeze front.

What affects the sea breeze?

The Coriolis effect, which is caused by the spin of the earth and described in **Buys Ballot's law**, has a significant impact on the sea breeze. We are accustomed to seeing this on the large scale with high and low pressure areas on synoptic charts but it also occurs with local sea breezes. In the absence of other effects, the 'classical' sea breeze will set in directly onshore but will then veer. When well established, it will blow more nearly along the coast. This 'classical' textbook sea breeze is most likely to occur along a coastline with no significant headlands and bays.

On a typical sea breeze day, the sea breeze will start around mid-morning with a light onshore wind. Over the next two to three hours, it will veer and increase in strength; by early afternoon it will be nearly parallel to the coast. On the eastern coast of Sardinia, shown in Fig 3.4, and other Mediterranean coasts, the sea breeze sets in about 1000 local time and can easily reach force 7 by 1300.

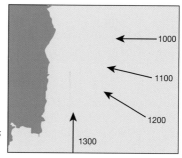

Fig 3.4 *Sea breezes usually veer*

What affects the sea breeze?

The ways in which sea breezes form and behave vary from place to place and there are no general all-embracing descriptions. These examples of sea breezes will give an indication of what to look out for, though.

Topography can often shape the sea breeze so that it becomes unexpectedly strong or blows in directions that may seem strange. The sea breeze cannot blow over cliffs or steep sloping ground but will be deflected around them, often with an increase in strength. A good example is around Cabo San Antonio on the eastern coast of Spain, opposite Ibiza. Force 7 or 8 winds are not unusual with a sea breeze here.

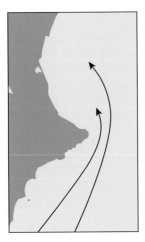

Fig 3.5 *Headland bending a sea breeze*

A similar effect can be found in the strengthening of the wind around Berry Head on the south coast of England. Here, the main driver of the sea breeze is the strong heating of the towns of Torquay and Paignton. Anyone who sails regularly between Dartmouth and

What affects the sea breeze?

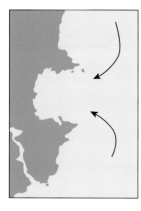

Fig 3.6 *Tor Bay can do this to a sea breeze*

Torbay will be aware of the strength of the south-south-west head wind during the afternoon. It can be a hard beat back to Dartmouth on a sea breeze day.

The Torbay sea breeze is complicated by the high ground to the north of the bay. The heating of the two towns can be so strong it can 'pull' the air from the north into the bay. The sight of two yachts running under spinnaker towards each other is well known here. As they approach each other, the wind will become variable and then each will start beating in opposite directions. The same effect can occur outside Plymouth Sound with Rame Head in one direction, and Yealm Head and the Great Mewstone in the other.

For rather different reasons, the Isle of Wight is another example of directly opposing winds being encountered in the same locality because of the sea breeze being shaped by topography. The strong heating of Southampton and nearby conurbations draws the sea breeze around both ends of the Isle of Wight and into both arms of the Solent.

What affects the sea breeze?

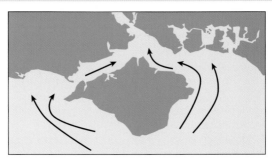

Fig 3.7 *A common sea breeze effect around the Isle of Wight*

The Solent provides a good example of what local heating can do to the wind. The sea breeze effect can be strong enough to reverse a north-east force 3 morning wind in the western Solent into a south-westerly force 4 by early afternoon. Heating of the Isle of Wight might be expected to create easterly winds along its northern side. However, the mainland heating is much stronger and the island heating effect may just lead to a more variable and lighter wind than on its northern side.

There are other examples of sea breeze effects interacting with each other. During the morning, a fairly weak sea breeze can set into the St Helier Bay on the south of the island of Jersey. Later in the day, the heating over mainland France becomes dominant; the St Helier sea breeze weakens and gets reversed as the St Malo area sea breeze takes over.

For similar reasons, the sea breeze along the south coast of Spain can be surprisingly weak. Although the gently sloping ground backed by hills facing the sun look ideal for a sea breeze, the effect of North Africa can dominate. A promising day's sea breeze sail can turn into a disappointing motor-sail.

Land breezes

Cooling over land at night leads to higher pressure than over the sea and therefore a flow from land to sea. However, the effect is usually less well-marked and land breezes are generally lighter than sea breezes. The effect can be enhanced by the air being channelled by river valleys and even more so by the steep sides of the Scottish lochs and Norwegian fjords. These katabatic winds, as they are called, can be strong, up to gale force, in such areas and particularly around Greek islands, the Adriatic and the Aegean.

In Fig 3.8, the boat anchored with an on-shore breeze of force 3 to 4, as can be seen by the waves on the water.

After sunset, this died away, the land cooled and during the night the wind blew in the opposite direction as it came down the valley at a steady force 3 for several hours. It helps to bear such effects in mind when you are in a crowded anchorage or near moorings and trying to decide how your boat and the others will swing.

Fig 3.8 *This yacht faced the shore overnight*

Sea/land breeze cycles

Sometimes, the sea breeze by day and the land breeze at night can take the wind round in an apparently near-continuous cycle. This is a well-known effect in the Belle-Île, Quiberon area of western France, as shown in Fig 3.9. The typical situation is when there is high pressure to the north giving a pressure **gradient wind** from the east or north-east.

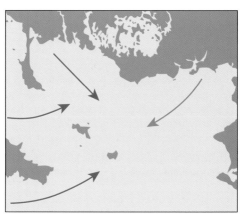

Fig 3.9 *A sea/land breeze cycle*

The night-time land breeze might come down the Vilaine River or out of the Golfe du Morbihan. The sea breeze counteracts that and a light south-west wind develops. This veers to become north-west. With night-time cooling, the land breeze sets in again. It is worthwhile bearing this in mind when trying to decide which anchorage to use. One that seems to be exposed to the daytime sea breeze is likely to be sheltered from the night-time force 4 breeze off the land.

Sea/land breeze cycles

SEA AND LAND EFFECTS

A rather similar effect occurs on the north-east of Mallorca, shown in Fig 3.11. Here, the slope of the ground around the bays of Alcúdia and Pollensa result in a steady force 3–4 overnight and during the early morning.

As the land warms up, the offshore wind dies away and, by about midday, the sea breeze sets in to both bays. This can give a good sail out of one bay during the morning and into the other in the afternoon.

Fig 3.10 *Hills surrounding the bay of Alcudia, Mallorca*

Fig 3.11 *Early morning offshore (land) breezes and afternoon onshore (sea) breezes*

Sea/land breeze cycles

The sea breezes described here are based on observations when the early morning wind was either calm or a light land breeze was blowing. Sea-breeze-type effects may well occur in other situations when there is already a wind. If there is enough heating to warm the ground, then the pressure pattern can be modified just due to the heating.

How strong? How far out to sea? How far inland?

The answers to these questions depend on where you are, how hot the land can be by day and how cool at night. Here are a few pointers.

◆ A well-developed sea breeze can reach force 4 or 5 around the UK but stronger when forced around headlands.

◆ In warmer areas, winds up to force 7 can easily be achieved or exceeded due to headland effects.

◆ A typical English south coast sea breeze can extend some 12 miles out to sea.

◆ The English south coast sea breeze can reach Salisbury, 30 miles inland, by evening.

In extreme cases, the sea breeze effect can be felt halfway across the English Channel; the effect can be felt 30 miles out to the south of the Isle of Wight and about 40 miles south of Lyme Bay. This is clearly illustrated in Fig 3.12, a detailed analysis of wind directions of the sea breeze on to English and French Channel coasts on an exceptionally hot day (reproduced courtesy of the UK Met Office). The solid black line shows the region where the sea breeze divides towards the English south coast and towards the French coast. In hotter regions of the

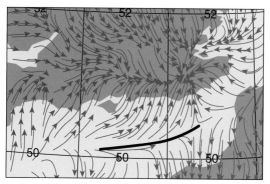

Fig 3.12 *Sea breezes can extend out to mid-Channel*

world, sea breeze effects may reach up to 100 miles out to sea.

Cliffs, straits and headlands

In addition to the heating or cooling of the land, there are the 'mechanical' effects of headlands, cliffs, straits and shorelines in general, all causing changes in both wind speed and direction. Here are some of the effects to look for.

Wind blowing towards a line of cliff will be forced to change direction and move parallel to it, as shown in

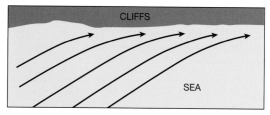

Fig 3.13 *Why the wind heads when approaching a coast*

Cliffs, straits and headlands

Fig 3.13. There is likely to be an increase in the speed as well as directional changes. The effect of being headed as a yacht approaches the coast is well known, and this is one reason.

An offshore wind coming over a cliff may give eddy effects, as shown in Fig 3.14. The 'splashdown' of the wind can be an area of stronger winds; look for a more disturbed sea as far out as ten times the height of the cliffs.

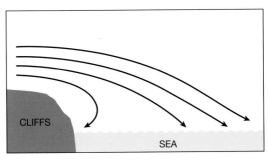

Fig 3.14 *Winds offshore*

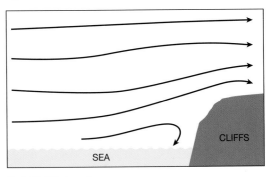

Fig 3.15 *Winds onshore*

A wind directly onshore to a cliff may have an effect like the one illustrated in Fig 3.15. In a wind directly onshore, local sailors have been known to anchor close to the Babbacombe cliffs using the eddying effect.

Wind blowing through a strait will be squeezed and thus increased in strength. Once through the strait, the wind will fan out. Through the Dover Strait, winds can be about one force stronger than up- or downwind. Through the Gibraltar Strait, the effect is more marked, while through the Strait of Bonifacio between Corsica and Sardinia, the effect can be an increase of four forces. In other words, a north-west force 5 upwind of the Strait can be accompanied by a force 8 or 9 through the Strait itself. A similar effect occurs in the North Channel between Northern Ireland and the Western Isles of Scotland.

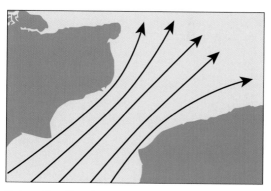

Fig 3.16 *The effect of straits on winds*

Cliffs, straits and headlands

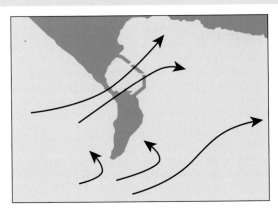

Fig 3.17 *Complex topography has complex effects*

Around Portland Bill winds will form eddies, and be squeezed between the Bill and the mainland. Changes in direction as the wind fans out over Weymouth Bay are well known and used by dinghy sailors.

What happens on specific days will depend on the direction and strength of the wind approaching the Bill. Local knowledge and careful observation will pay handsome dividends.

Summary

Like much of meteorology, little is clear cut, and simplistic descriptions or explanations are rarely sufficient or complete. Remember that heating leads to lower pressure, and cooling leads to higher pressure. Think about Buys Ballot and it should start to make sense. Imagine the land as a low pressure area and, if it is an island, then there could possibly be a sea breeze effect all round. Forecasts that are claimed to be precise can rarely be so because of all the various local effects, as well as the problem of predicting such imponderables as the amount of cloud. GMDSS forecasts may just say 'sea breezes' or, in French 'brises'. After that, it is up to you!

Watch what happens, watch other boats; learn by your own experience. Listen to local 'experts' but bear in mind that memories are selective. Something said to 'always' happen rarely does so! Try to distinguish between the effects of heating or cooling, mechanical effects of topography on the wind, and other effects such as gusts near showers and larger-scale effects such as near fronts. It will not be easy but you will learn. That is the story of weather – and sailing.

Understanding sea state

Particularly to a small boat, the state of sea can be more critical than the strength of the wind. Waves and swell occur due to a combination of wind, tide or current, and bottom and coastal topography. Only by appreciating how and why these factors affect the sea state can the sailor be fully prepared.

Wind waves

A light wind over water is affected by friction, resulting in small-scale turbulence; this leads to ripples on the water surface. As the wind increases, the ripples get larger to become wavelets. Further increase in the wind will increase the size of the waves, which start to be pushed along by the wind.

The waves travel slower than the wind so that the force of the wind makes them grow until they become too high relative to their wavelengths. The ratio of wave height to wavelength is a measure of the steepness of waves – when this is about 0.15, waves start to break

Fig 4.1 *A Force 5 kicks up a sea*

and white crests appear. Typically, this is around force 3 or 4. Fig 4.1 shows waves in a force 4–5.

Wind speed and direction vary continuously so that there is always a range of wave heights and wavelengths. The speed at which waves move increases with wavelength so that, as the wind increases, the ranges of wave heights and speed also increase and the sea becomes increasingly confused.

Almanacs such as *Reeds* have pictures of sea state for winds up to force 12. It is good practice to keep monitoring the sea state as additional information on wind strength. This is an aid when it comes to interpreting a forecast in terms of the effect of the wind on the sea as well as on your boat.

Swell

Wind waves travel downwind, usually in groups, with the bigger, longer waves travelling faster than the smaller ones in the group. As they reach the front of a group, these faster-moving waves disappear but the group as a whole continues to move. As groups of waves move out of the area where they started, they become known as swell. Having lost the biggest waves, the steepness decreases – so the distinguishing characteristic of swell waves is that they do not break.

> Because swell wave groups move well away from the area of their formation, they provide an indication of what is happening over the horizon. Particularly for offshore sailors, a long swell will often be a good indication of stronger winds approaching and of the direction of those winds.

Wind waves

Fig 4.2 *A Force 10 is best avoided!* © Tony Cross

Waves of differing wavelengths and heights move at different speeds, so waves can come into and out of phase. This leads to spells of higher waves and spells of smaller ones. The old saying about every seventh wave being a big one has some slight statistical justification. The effect, as seen by the sailor, is of sea conditions that are varying continuously, and particularly so when large, locally generated wind waves are superimposed on the swell. The seas shown in Fig 4.2 were experienced in a force 10 in the English Channel.

Water depth and tidal stream

As swell moves into shallower water, the effect of the rising sea bed is to restrict the maximum wave amplitude. The energy in the waves does not change and is realised in a decrease in wavelength, giving an increase in steepness. This may well be manifested as breaking water.

A similar effect occurs when swell meets an opposing current or tidal stream. The wavelength is reduced, leading to the steepness increasing. In both cases a rougher sea results, especially if the wind is strong and wind waves are large.

The reverse happens when the tidal stream changes to be in the same direction as the swell, or when a swell propagates into deeper water; in both cases, the result is an increase in wavelength and a lower steepness.

Reflection and refraction

Waves approaching or near to a coast can be affected in various ways. Reflection is most likely when waves meet a solid wall, such as steep cliffs rising out of the sea or a harbour wall. Fig 4.3 shows the effect in very shallow water, with waves coming from one direction being reflected to a totally different direction. On a large scale, this can be a serious hazard near the entrance to some ports. Dover is one example where there are substantial walls exposed to a big swell from the west.

Fig 4.3 *Waves can be reflected on all scales*

Reflection and refraction

Fig 4.4 *Shallowing water can have alarming effects!*
© Alan Lapworth

Refraction occurs, typically, when a swell is moving in a direction parallel to a coast. The effect is similar to a line of soldiers being ordered to 'wheel'. Those at one end of the line take much shorter steps than those at the other end and the whole line rotates. Inshore, the shallower water results in steeper waves that move more slowly than those further out. These continue at their normal speed and the swell train rotates. The effect can be seen on a gently sloping beach, where waves are usually seen to come onto the beach directly even though the wind is along the coast.

The refraction effect can be marked around headlands such as Start Point in South Devon or the ends of the Isle of Man. Refraction can make a seemingly pleasant anchorage most uncomfortable and this seems to be a feature of many Mediterranean bays. The typical scenario is that of entering a bay with a moderate wind and assuming that as the wind dies away, so will the sea.

However, as the wind dies, the wind waves disappear but swell waves work their way in around the headlands to become more noticeable. There may also be some reflection off the sides of the bay. A very disturbed night can be the result.

Sea state – general advice

Tidal stream, water depth, reflection and refraction occur together to varying degrees and always have to be borne in mind when approaching a coast or entering a river or a port. Some harbour entrances are renowned for being difficult – Chichester and Langstone harbours, Salcombe and the river Étel in western France are some that are best avoided with an ebb tide and any swell. Navigation charts sometimes show where tidal stream and uneven sea bottom effects create dangers. Such warnings, often labelled 'overfalls', should always be heeded.

> Swell can be deceptive, particularly when it is a long way from the initial strong winds causing it, and a big swell may be present with light winds. The smooth rise and fall of the sea might look innocuous until it moves into shallowing water or a contrary stream, when the steepness may increase alarmingly. When near a hazard such as a harbour entrance, it may suddenly become too late to take avoiding action.

Many European sailors will be aware of the effects of swell on certain harbours and rivers around Biscay coasts and the Iberian Peninsula. Pilot books and almanacs usually provide good advice. If in any doubt, call the nearest MRCC (they should all speak English) and ask. They would rather give advice than have to mount a rescue operation later. Warnings may be broadcast on VHF.

Lagoon, bay and seiche effects

If the wind is strong enough, frictional drag can cause water to be pushed to the lee side of a lake, lagoon, bay or sea. This can be seen on a wide range of scales from Lough Neagh in Northern Ireland, Adriatic bays and lagoons, the Baltic Sea and the North Sea right up to the oceanic scale. The effect across the North Atlantic in a strong south-westerly can be a precursor of North Sea tidal surges.

When the wind eases or changes direction, the water tries to return to its normal level; but the momentum can cause an overshooting effect and the water piles up on what was the windward side. The same process then occurs in the opposite direction with a similar overshoot. This will be repeated with decreasing amplitude.

This effect is enhanced when the water can only escape slowly from a lagoon with a narrow entrance. The period of the oscillation can be a matter of tens of minutes for a small body of water, to hours for a large lagoon, to days for a sea. In lagoons, the amplitude can be up to a metre across a lagoon and some tens of centimetres for the Baltic.

Other seiche effects seem to be caused by topography in certain specific directions of swell. Well-known examples include the harbours of Port Tudy (Île de Groix, Brittany) and Ciudadela (Menorca), where rises and falls of water level of up to a metre can occur over several hours. There may be warnings of these events on the VHF.

Fig 4.5 *Lagoons with narrow entrances, like this one in Venice, are prone to seiches*

Tsunamis

These are caused by seismic events such as underwater earthquakes. In principle, they can occur anywhere, although areas near faults in the earth's crust are especially prone. First signs of a tsunami are a withdrawal of water from the shore before the surge comes. Like all swell effects, there will be a fairly small and barely noticeable wave out at sea but greatly increased height as it approaches shallow water.

There is little that the sailor in harbour or at anchor can do. Experiences of a fairly small tsunami affecting Mahòn, Menorca, and Torrevieja, Spain were that boats in marinas suffered by being caught under pontoons because they were moored too close. Also, some boats at anchor or on moorings hit the bottom in the withdrawal of water and broke free as the surge came in.

Why weather prediction is so difficult

Weather forecasting is, in some ways, a paradox. Anyone capable of grasping the basics of school level physics can understand the factors that determine the workings of the atmosphere. The complications come in the interactions between the many simple processes – for example, those shown in Fig 5.1 involve heating and cooling of the air.

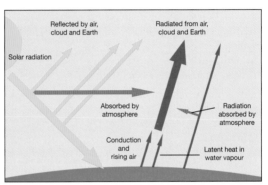

Fig 5.1 *Radiation, conduction, convection and latent heat all play their part*

The important processes are:

◆ The sun heating the earth by radiation.

◆ The earth heating or cooling the atmosphere by:
  • Conduction.
  • Convection.
  • Radiation.
  • Latent heat effects.

◆ Heating differences creating temperature differences.

◆ Temperature differences creating pressure gradients.

- ◆ Pressure gradients making the air move.

- ◆ The spin of the earth – the Coriolis effect.

- ◆ The effects of topography including land use.

Feedback between all these processes results in weather being far more complex than many people realise. This is true on all scales – gusts, showers, sea breezes, larger scale lows and highs, right up to the largest scale of climate change.

As sailors, we see at first hand variations in weather, sometimes quite marked, over small distances and short times. Isaac Newton had the answer when he said that applying a force creates acceleration or change. There are always forces in the atmosphere – pressure gradients, gravity, and the Coriolis effect. Therefore the atmosphere is always in a state of flux. Also, the effects described above lead to continual changes in the pressure gradient. The end result is that weather forecasting is rather like trying to hit a moving target.

For more complete theory, there are many meteorological textbooks with varying degrees of detail, simplification and complexity. Anybody wishing to know about the basics, such as the Coriolis effect, can visit my weather website.

Before computers

Forecasting used to be based on what is termed 'synoptic meteorology'. The starting point was the subjective analysis of weather data at various heights, including the jet stream and above. Forecasters then used a combination of theoretical ideas and collective experience. Although there was some success, the ultimate stumbling block was the inability of the human

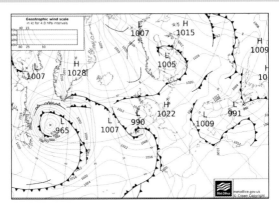

Fig 5.2 *UK Met Office charts are often complex*

brain to quantify the theory and assess all the available data. Skill was strictly limited.

Synoptic charts are still produced by the UK Met Office and other National Meteorological Services (NMS). They are issued as analyses, as shown in Fig 5.2, and forecasts are given in a kind of meteorological shorthand. The isobars provide information about wind strength and direction, while the fronts tell you something about the weather.

Numerical Weather Prediction (NWP)

The physical processes driving the atmosphere, shown diagrammatically in Fig 5.2, can all be expressed in terms of mathematical equations and are referred to as 'weather models'. The processing of these models by the most powerful computers available is known as NWP. This is the basis of all marine weather forecasts. These computer model forecasts are usually produced on a six-hourly cycle using as much observational

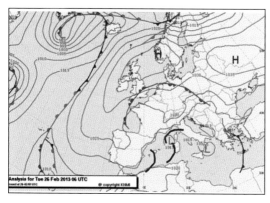

Fig 5.3 *Dutch charts may be easier to use than UK ones*

information as the computers can handle. Forecast data up to eight days ahead can be made available a few hours after each of the four base times of 00h, 06h, 12h and 18h UTC.

In itself, all that may not matter greatly to the user of shipping or inshore waters forecasts. However, some background knowledge will help in understanding what forecasts can and, more importantly, cannot do. This is most necessary because sailors can now obtain both raw and processed output from NWP models, which can be downloaded to a PC or tablet. Effective use of these and all other weather forecasts depends on having a little knowledge of their strengths and weaknesses.

The data is available in Gridded Binary (GRIB) files, a data format devised by the World Meteorological Organization (WMO) to enable the international exchange of data and its storage. The binary format compresses the data so that a small file can be translated into a large amount of information.

Displays of wind vectors at three- or six-hourly intervals for the next eight days, like this screenshot from a tablet app (see Fig 5.4), may look convincing but they must be used with care.

In the same way that a medical prognosis depends upon a good diagnosis, any weather forecast will

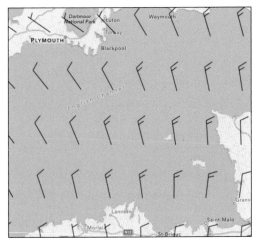

Fig 5.4 *Sailors will find NWP output clearer than synoptic charts*

depend upon having a good analysis of all the available observational data. That is even more true now than it was in the days of synoptic meteorology, when much of the available information could not be used effectively.

In order to be able to predict the weather anywhere for more than a few hours ahead, it is necessary to know about the weather everywhere. This is because of all the many interactions. The starting point for all NWP calculations is an analysis of as much of the available data – over the whole globe and up to heights of about 80km – as the computers can handle.

The data used in modern weather forecasting comes from several different sources:

◆ Manned and automatic weather stations on land.

◆ Ships and data buoys, tethered and drifting.

◆ Free flying (radiosonde) balloons.

◆ Aircraft.

◆ Satellites in low earth and geostationary orbits.

Limitations to NWP

Since its introduction, NWP has been and will continue to be under continual development, with more or less steady improvement. There are three major factors that limit predictability and that the scientists are addressing.

◆ First, the various observing systems have different characteristics in terms of reliability, accuracy, coverage and in the ways that they represent the atmosphere. Their assimilation into a coherent pattern requires as many computing resources as the forecast itself. There is no unique way in which the data can be analysed and there is always some uncertainty in the starting point for a forecast.

◆ Secondly, even if the data analysis were perfect, there are many interactions between the processes driving the atmosphere to consider. The atmosphere is like an analogue computer whose power greatly exceeds that of any digital computers so far envisaged and, possibly, of any that will ever be built.

Numerical Weather Prediction (NWP)

◆ Thirdly, there is chaos on all scales – from the small eddies that blow paper around our streets right up to the vortices, thousands of miles across, shown on satellite images. In between, there are convective weather systems such as cumulus clouds and their big brothers, cumulonimbus, producing showers and thunderstorms.

On the **synoptic scale**, major storms can and do develop from small beginnings. Hurricanes start as small groups of showers or thunderstorms; many of the frontal depressions that reach western Europe start as minor features off the eastern seaboard of the USA and are not always discernible on a satellite picture.

Weather predictability is limited by:

◆ Small-scale data uncertainties.

◆ Computer power.

◆ Chaos.

Ensembles

You may have heard and seen the term '**ensemble** forecasting' on, for example, the WeatherOnline website. The technique is used by the major forecast centres, including the European Centre for Medium-Range Weather Forecasts (ECMWF), to counter the data uncertainty problem. After a run of the NWP program suite, a diagnostic program shows areas where small data analysis uncertainties might have a significant effect on the outcome in a particular area.

Small random variations in the analyses can then be introduced while maintaining consistency with the observational data. The forecast is then run many times – 50 in the case of ECMWF and 24 for the UK Met Office (in 2013). This allows forecasters to get an idea of how certain the forecast is for their area of interest on any particular day.

The results can be seen when TV or radio forecasters are sometimes positive about a forecast for the next week and at other times much less sure. It is a moot point whether or how this uncertainty could or should be used in marine forecasts.

Grid spacing

One factor to bear in mind when using GRIB data is the effect of calculations having to be made on a grid of points. The UK Met Office uses a three-dimensional grid with a horizontal spacing (in 2013) of about 25km. This limits the detail that can be analysed, as is shown in Fig 5.5 by placing a 25km grid over the Isle of Man – roughly 30km long by about 10–15km wide. As the figure shows, the island could slip totally between the

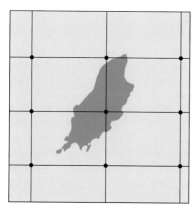

Fig 5.5 *A 25km grid over the Isle of Man*

computer grid points. At best, there can only be two points over the island. The same would be true for weather detail of comparable size.

> Any grid can only represent detail on a scale of about four or five grid lengths – 100–125km+ – for most global models run by the major operational forecast centres. To prevent the mathematics from generating spurious detail, the models filter out detail of anything less than this size. The result is a smoothing effect, so the strongest winds are not well represented. The same is true of small detail and the effects of small-scale topography.

Small-scale NWP

To help prediction of detail, such as localised heavy rain or snow, especially when due to topographic effects, national weather services use limited area (mesoscale) models with short grid lengths. The UK Met Office mesoscale model (in 2013) has a grid spacing of 1.5km, with weather and topographic data input on that scale.

The area boundaries have to be continuously updated from the much lower resolution 25km global model. This means that movement on the synoptic scale can limit the prediction for small-scale detail to only a few hours ahead. Another problem is convection which, as can be seen on satellite images, is random. Individual clouds have short lifetimes. While a group of showers will have a significant lifetime and be predictable, the individual elements are short-lived.

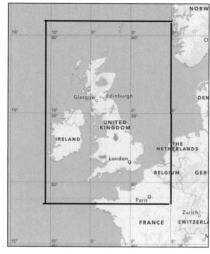

Fig 5.6 *Detailed forecasts only cover small areas*

As with all forecasts, care has to be exercised in the use of these detailed models. Although they will improve, remember that usefulness might currently be affected by:

◆ Lack of detailed information outside the forecast area.

◆ Convection, which is random and not predictable in detail beyond the first few hours.

◆ The lifetime of individual convective cloud, which is a few hours maximum.

◆ The generally short lifetimes of small weather details.

Numerical Weather Prediction (NWP)

Fig 5.7 *IBM super-computer (photo courtesy of ECMWF)*

There is little doubt that accurate – or, at least, useful – short period forecasts can be produced for a few hours ahead and tailored to a specific requirement, such as a regatta. However, the lead time between data collection, running the computer model and disseminating the forecast is such that a broadcast service to meet general needs is unlikely in the foreseeable future. National weather services optimise the use of these detailed models, for such purposes as heavy rain or snow warnings, by running them as ensembles.

Several sources of these detailed forecasts are available, some on prepayment and some free. Some services are listed in the Appendix.

Summary

When using any computer forecast, remember that:

◆ Any model can only analyse weather data on a scale of about five grid lengths.

◆ This has implications for the prediction of strong winds and topographic effects.

◆ Large-scale patterns will often, but not certainly, be well predicted for up to seven or eight days ahead.

◆ The lead time between data acquisition, processing into a forecast and distribution limits the value of mesoscale forecasts.

◆ Mesoscale models can predict convection and some other relatively small effects such as local sea breezes in general but not specific terms.

For further reading, the UK Met Office and similar websites in other countries have informative pages on NWP, including mesoscale and ensemble forecasts. In addition, Huw Davies has written *Numerical Weather Prediction – A practical guide for mariners* to help all sailors have a greater understanding of NWP and the sources of GRIB files than is possible to provide here.

Planning

It may sound rather grand to talk about a passage plan for a short trip that you have done many times before or just an hour or two day sailing off the mooring, marina berth or beach and back. Nevertheless, whenever going to sea, it is sensible to assess expected weather conditions and any other factors that may affect safety.

The Marine and Coastguard Agency (MCA) says it is a legal requirement to make a passage plan in proportion to the trip to being undertaken and that this should include, as appropriate:

◆ Taking account of weather forecasts.

◆ Getting regular updates while at sea.

◆ Using tidal predictions.

◆ Assessing suitability of boat, equipment and provisioning.

◆ Noting possible navigational dangers.

◆ Making contingency plans.

All these points involve knowing about weather and any possible effects of weather that might result in danger.

Weather and whether to go?

The big question for all leisure sailors, with or without any formal training in meteorology, is how to use inevitably imprecise weather forecasts. In particular, the question is how to assess the reliability of a forecast, bearing in mind that there are limits to predictability, especially of short term local detail. For the large-scale weather pattern, it is also clear that reliability will decrease with time ahead – but by how much?

One approach is to take forecasts at face value – but given the uncertainties and limitations in a worded, necessarily brief text or a forecast synoptic chart, that is not a sensible way forward. Many complaints about inshore and offshore forecasts are because users try to treat a general forecast as being applicable to the whole of an area for the next 24 hours.

To get the best out of forecasts, it is necessary to apply judgement and experience; these only come with time. This section will point the way to use weather information rather than give a detailed route map. After that, it is up to the sailor to learn by experience – as is the case for most aspects of sailing.

Day sailing and coastal passages

The first priority is to know what the GMDSS inshore and offshore forecasts are saying. Remember that any forecast text for a long stretch of coast or a sea area will be little more than a brief summary. However, they should give warning of hazards.

> Monitor the forecasts for at least 24 hours before going to sea to ensure you have heard the forecast correctly. Listen carefully to date and time of issue; these are the vital pieces of information in any marine forecast and without them the forecast is meaningless.

It is clearly necessary to understand the terminology. Some of the words listed in the Appendix are universal, some refer specifically to the UK Met Office.

Listen carefully for changes in the forecasts, especially trends. Listen to successive forecasts to see if there are changes in the timing of, say, strong winds or their

Day sailing and coastal passages

Berwick upon Tweed to Whitby

Strong winds are forecast

For coastal areas up to 12 miles offshore from 0600 UTC Fri 24 May until 0600 UTC Sat 25 May

24 hour forecast:

Wind North or northeast 6 to gale 8, decreasing 4 or 5, decreasing 3 or 4 later.

Sea State Very rough, becoming moderate.

Weather Rain or showers, fair later.

Visibility Moderate or good.

Outlook for the following 24 hours:

Wind North or northeast 3 or 4, becoming variable then northwest 3 or less.

Sea State Moderate, becoming mainly slight.

Weather Fair.

Visibility Good.

Fig 6.1 *Texts of VHF broadcasts, like this one from HMCG, can often be found on the internet*

strengths. Inconsistencies between successive forecasts imply uncertainty and mean that you should be rather more careful. National weather services keep forecasts under continual review as new data arrives and as their computers produce new predictions. Some countries – the UK, Ireland and the USA among them – update forecasts every six hours; others less frequently.

Secondly, use GRIB files and synoptic charts whenever available to give a better idea of variations in the overall pattern in space and time. Neither can show small spatial detail very well but the GRIB output is the more useful, particularly because it will give a better indication of how the pattern is expected to change in time. We will be looking at using GRIBs later.

When out sailing, keep watching the weather to check that it is doing what was expected. If not, then look back at the forecast to confirm your understanding. Make sure you hear the next one. If you hear a forecast or warning on the VHF and do not understand, especially when abroad, then call up on VHF and ask. Coastguards are there to help; they are not meteorologists but they can ensure that you have heard the message correctly. If a change has been forecast, look for signs such as increasing cloud, the wind starting to back or veer or the barometer starting to fall or rise.

> In addition to the forecast, use your own experience of local effects. If you sail regularly in an area, you will know how the sea breeze behaves and what the nearby headlands and estuaries do to the wind. If you do not know the area, try to apply your general experience coupled with the descriptions of local effects in this book. You may not get it right but you will be better prepared for local variations, most of which will not get mentioned in a typical GMDSS forecast.

Passage making

It may seem a paradox, but coastal sailors can be more at risk than those crossing oceans. Even on short coastal passages, a slow-moving yacht can be many hours from an accessible port of refuge. Many motor boats, particularly those with planing hulls, will have their speeds reduced in a moderate sea likely with a wind of force 4 to 5. Even quite large motor boats will have greatly reduced speeds in a rough sea, say with a force 6 and above. Coastal sailing yachts and many motor boats are unlikely to be equipped as well as a blue-water yacht. In any case, there may well not be enough sea room to ride out a storm.

ER

AST

Passage making

Perhaps more importantly, many English coastal sailors will be cruising for fairly short periods of a week or two, with the imperative to get the boat back to their home port or charter company. They may have to make a difficult choice between leaving the vessel in a distant port or going to sea in unfavourable conditions.

Having tight deadlines can, all too easily, become a recipe for bad decision-making. Successful cruising, whether in a power boat or a yacht, is all about looking ahead to avoid being in the wrong place at the wrong time. In sailing terms, that means avoiding being in a port you do not want to be in, in weather you do not want to go out in.

With or without tight deadlines, it is sensible to use the advances in weather forecasting to anticipate problems and make decisions that maximise enjoyment and minimise risk. You can use synoptic charts for this purpose but GRIB file information is easier to use and enables you to think up to a week ahead.

To make optimum use of GRIBs, study them daily for the next eight days and ask the following questions:

◆ Is there a weather window for my next passage?

◆ Is a window opening or closing?

◆ Is there a suitable alternative?

◆ Is adverse weather expected?

◆ If so, what are the options?

By looking at synoptic charts and GRIB forecasts on at least a daily basis and by monitoring GMDSS forecasts, you will be keeping up to date with the weather and have the best idea possible of what might happen. Having decided that the weather is suitable to go tomorrow, it will be unlikely that the GMDSS forecasts

92

on the day of departure will be a surprise. Keep checking, nevertheless! Final checks on the day are to consider:

◆ Is the most recent forecast still favourable?

◆ Must we go?

◆ Is there a better option, such as staying in port?

◆ Are we going to the right place?

For example, leaving an English Channel port bound for South Brittany, it might seem reasonable to stop for a night or two at l'Aberwrac'h on the north coast of France. However, looking at the forecast for the following few days might suggest that you could become weatherbound there with strong westerly winds. It might be a better option to make a longer passage, say, to Camaret or even further south.

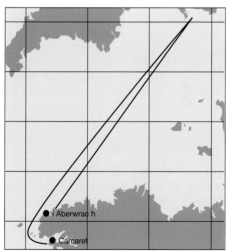

Fig 6.2 *Consider your passage options in advance*

Passage making

Although forecasting has improved greatly, there will always be some uncertainty and it is useful to have some idea about how reliable the outlook is for the next few days. The approach to this is similar to that when using forecasts for the next 24 hours:

◆ Check each forecast against its predecessors.

◆ Contrast and compare the outcomes.

◆ Look for consistency.

◆ Look for trends.

◆ Make plans.

◆ Above all, keep your plans updated.

That may seem easier said than done, but here are two examples of forecasts for the next week, showing when planning ahead is possible and when it is not.

Example 1: Certainty

A typical decision was that faced by a rally organiser hoping to finish up in Étel, western France, a port noted for its difficult entrance in strong onshore winds. Participants wanted to be sure of being able to depart after the rally, on 30 June. On the basis of the forecasts on 23, 24 and 25 June, all for 30 June (see Fig 6.3), it was clear that strong westerly winds were most likely. Plans were changed accordingly.

Example 2: Uncertainty

Forecasts for the next week are not always as clear cut as the previous example suggests. The set of forecasts in Fig 6.4 for the Vendée–Charente–Gironde area shows forecasts on four successive days all verifying on the same day, 23 September. The differences from one day to the next were a clear warning that, on this occasion, the forecasts could not be used for planning purposes.

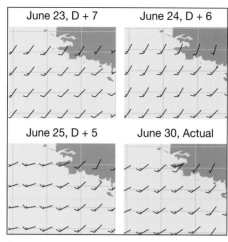

Fig 6.3 *Consistency inspires confidence*

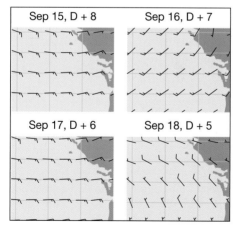

Fig 6.4 *Inconsistency implies uncertainty*

Although such different forecasts might not seem very helpful, the negative result has value in providing an indication of uncertainty. The idea behind this approach is that forecasts for the same weather situation are being generated using different data inputs. Think of it as a kind of ensemble forecast to show whether the weather situation is one that is reasonably predictable over the next few days or not. It is a way of minimising the effects of chaos.

Learning by experience

You can gain confidence in using GRIB data in two ways. First, go through the kind of exercise shown here. Look at eight-day, seven-day, six-day, five-day forecasts on successive days. When they show consistency, verify them against the analysis for the day to which they refer, i.e. the T=0 'forecast' for that day. That should give an idea of how to judge, on any particular occasion, how reliable the forecasts might be for several days ahead.

Secondly, use a GRIB forecast to look at each of the next eight days. Then, verify each forecast against the analysis for each of the relevant days. That will give some idea of how forecasts become less reliable the further ahead they are predicting.

Repeat these exercises several times – five, at least – to get an idea of how useful the forecasts would have been for decision-making in the area where you would be sailing.

When using GRIB forecasts, it might be sensible to check at 12-hourly rather than daily intervals – particularly within about three days of the intended departure date – if forecast conditions look marginal, say in wind direction or strength. NWP computers usually run four times a day using data with nominal data times, i.e. T=0, of 00, 06, 12 and 18 hours UTC. The runs based on 00 and 12 hours UTC are likely to be more reliable than the other two, because more in situ data is available at 00 and 12 UTC than at 06 and 18 UTC.

Ocean crossings

Small vessels with speeds of less than 10 knots cannot realistically avoid major storms by weather routeing on long passages. Storms are too large and forecasts become less reliable the longer the forecast period. After about six or seven days the reliability of forecasts decreases rapidly; what might have seemed a good decision for the first week might leave a yacht in a poor position for the second. Round-the-world racing yachts receive GRIB forecasts every six hours. They put the data into their computers, use weather routeing software, and revise their plans with each new set of data.

A cruising yacht can use weather routeing programs in much the same way. Decide which route you wish to follow, say, a great circle or a rhumb line to the next port or waypoint. Then, each day, use the latest GRIB files with a routeing program to get onto or keep on that track in, say, seven days' time. Update the course daily on that basis.

Alternatively, and this is what most will probably do, choose the best route and time on climatological grounds – for example, by avoiding crossing the Atlantic in the hurricane season. Then, leave port with a favourable weather window for at least three days to get clear of land. After that, keep a watch on GMDSS high seas forecasts, GRIBs and (if available) synoptic charts to get good warning of sustained bad weather. That should ensure enough time to prepare crew and vessel.

Summary

For all kinds of sailing:

◆ Do not assume that any forecast will be accurate.

◆ Look at every GMDSS forecast.

◆ Study GRIB data at least daily.

◆ Study synoptic charts – if available.

◆ Think and plan ahead whenever that is possible.

◆ Reduce uncertainty by looking for consistency in forecasts.

◆ Do this for GMDSS in the short term and GRIBs in the longer term.

◆ Treat inconsistency as an indication of uncertainty and plan accordingly.

Remember the old adage: 'There are old sailors; there are bold sailors; but there are few old bold sailors.'

Types of marine forecast available

There are three categories of forecasts available to sailors. First, text and synoptic charts in which a national weather service forecaster has added their experience in interpreting both weather and weather computer models. Both will necessarily be general in nature and brief when in text form.

Secondly, numerical weather prediction, with no added human intelligence or interpretation. Sometimes, this data comes direct from the computer; in other cases, there has been some presentational processing, but with no additional information or added value. Fine (meso-) scale numerical weather predictions are available from national weather services, international organisations or private firms – see the earlier chapter on Weather Forecasting (pages 76–86).

Consultants form a third group.

The Global Maritime Distress and Safety System (GMDSS)

All maritime nations participate in the GMDSS, which has two broad functions. One is to respond to emergencies at sea; the other is the provision of information relating to safety at sea. This includes the broadcasting of weather forecasts and warnings.

There are three categories of GMDSS forecasts in text with characteristics listed below.

For inshore or coastal sailing lasting a few hours, generally less than a day, and within 10–20 miles offshore. Broadcast twice to eight times daily.

◆ Updated at least twice daily.

◆ Warnings of strong winds, force 6 and above.

◆ In the national language.

◆ Sometimes available in English also.

◆ Local time (LT) is used.

For mainly short, offshore passages lasting from a few hours to a few days.

◆ Updated at least twice daily.

◆ Always available in English.

◆ Warnings of gales, force 8 and above.

◆ Times in UTC.

For ocean passages typically lasting weeks.

◆ Updated twice daily.

◆ Always in English.

◆ Warnings of storm force 10 and above.

◆ Times in UTC.

Pressures should be given in hPa; winds may be in Beaufort forces, knots or metres/second – to a good approximation, one metre/second = 2 knots. Although practice varies slightly, forecasts are usually valid for 24 hours with a – sometimes brief – 24-hour outlook. Some countries issue longer term outlooks either as extensions to the routine 24-hour forecast, or as separate broadcasts.

Forecast areas
Each country defines its own coastal areas, usually using well-known marine landmarks such as headlands, while offshore and high seas areas are agreed internationally. Offshore area names usually refer to well-known features found on navigation charts.

Fig 7.1 *Forecasts and warnings are the responsibility of agencies such as the MCA*

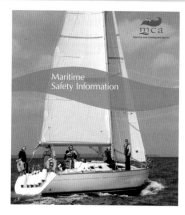

For organisational purposes, the world is divided, first into 21 METAREAs (identical to NAVAREAs), with one or more lead countries in each taking responsibility for coordination of the provision of forecasts.

Forecast texts

The main strength of these forecasts is that they are written by forecasters using the latest weather actuals and computer forecast. They benefit from human experience and judgement. In order to be clear and with tight limits on length, they are little more than brief summaries. They are required to mention potential hazards – strong winds, fog and thunderstorms. Although they should always be heeded in the interests of safety, they are not necessarily the most useful of forecasts when used in isolation.

An unintended consequence is that these types of forecast will slightly overestimate wind strengths, although the forecasters try not to do so. For UK forecasts, this is an effect – in part, at least – of having to issue warnings of gale force winds when they *may* occur but only being allowed to cancel a warning when it is certain that gales have ceased. A warning is a forecast; a cancellation is a statement of fact. Also, in the interests of brevity, forecasters are likely to write 'variable 3' when it would have been more accurate to say 'variable 0–3'.

Texts of forecasts will usually be found on national weather service websites. Before going to another country where the coastal forecasts may well not be available in your own language, it is advisable to use the internet to check on the words used, the form of the forecast, and any place names with which you might not be familiar. When at sea, the VHF radio may be the only means of receiving these forecasts and warnings; the ability to understand them could be vital to the safety of vessel and crew.

> Be aware that in some forecasts, particularly those broadcast by the UK, words such as 'wind', 'visibility', 'weather' and 'sea state' are often omitted in the interests of brevity without any loss of clarity.

Other GMDSS services
Information is available in the form of synoptic charts, actual reports and, occasionally, output directly from NWP models.

GRIB files and products
Numerical weather prediction output is presented in a compressed data form. This data is a valuable source of free information readily available to all.

Basic (free) services

These computer forecast outputs are usually from the United States' National Oceanic and Atmospheric Administration (NOAA) General Forecast System, usually referred to as the GFS. This is universally accessible and free of charge. An alternative source is the Canadian weather service, Environment Canada. GRIB data can be obtained directly by those with the necessary computer skills but most sailors use one or more of a number of free GRIB services. Data is (in 2013) available up to eight days ahead, at 0.5° latitude/longitude spacing, about 50km, and at three-hourly intervals. The output is slightly degraded from that available to the forecasters because the NOAA computer uses a spacing of nearer 27km.

Depending on the service used, it is possible to get a selection of wind speeds and directions, gust speeds, surface pressure, rain areas and amounts, cloud amount, air and sea temperature, sea state or swell and a number of other parameters including the thunderstorm index, CAPE (Convective Available Potential Energy).

You can obtain the basic information in two ways at no cost apart from those of communications:

◆ Direct file transfer to a PC or a tablet.

◆ By email attachment.

Direct file transfer is straightforward using software that is free for a PC or at a small charge for a tablet app. The use of an email service is a little more complicated but may be the only viable way if data network or WiFi services are not available. A small email attachment will often provide all that is necessary. The Appendix lists some of these services.

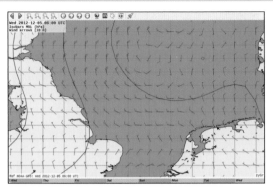

Fig 7.2 *Winds and isobars from GRIB files*

Fig 7.2 shows the output from a download to a laptop PC. The wind vectors follow the normal convention of one long barb = 10 knots, one half barb = 5 knots. The barbs point to low pressure. In this example, winds down the east coast of Britain are north to north-west. An eight-day forecast at six-hourly intervals for this area, received as an email attachment, was about 140 KB.

Fig 7.3 is a screenshot of a tablet app display of predicted wave and swell. Whether using a PC, tablet or email attachment service, the data is saved automatically for viewing later when offline. This is an important consideration when cruising and using some form of mobile connection.

Displays in chart form are a useful way of looking at forecasts for an area and to see the large-scale pattern. To help interpretation in an area of interest, time sequences can help. These can be in tabular form as shown in Fig 7.4.

Alternatively, they can be in graphical form as shown in Fig 7.5.

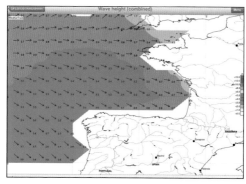

Fig 7.3 *Sea state from GRIB output*

	Thu 2013-01-24			Fri 2013-01-25				Sat 2013-01-26				
	06:00 UTC	12:00 UTC	18:00 UTC	00:00 UTC	06:00 UTC	12:00 UTC	18:00 UTC	00:00 UTC	06:00 UTC	12:00 UTC	18:00 UTC	00:00 UTC
d (10 m)	94° 9.9 kts 2 Bf	135° 13.5 kts 4 Bf	147° 13.7 kts 4 Bf	154° 20.6 kts 5 Bf	169° 24.4 kts 6 Bf	176° 28.6 kts 7 Bf	204° 26.5 kts 6 Bf	326° 14.9 kts 4 Bf	274° 16.9 kts 5 Bf	202° 25.3 kts 6 Bf	189° 17.5 kts 5 Bf	194° 32.7 kts 7 Bf
nd gust	10.1 kts	12.9 kts	13.7 kts	22.0 kts	27.2 kts	33.8 kts	33.0 kts	16.2 kts	5.7 kts	22.6 kts	36.7 kts	45.2 kts

Fig 7.4 *Forecasts at a point (using zyGrib)*

Tuesday, 26 February 2013	Wind	Pressure	Rain
00h	16.3 kts	1029.5 mb	0.1 mm/h
06h	21.8 kts	1028.0 mb	0.1 mm/h
12h	23.3 kts	1029.7 mb	0.0 mm/h
18h	19.4 kts	1031.4 mb	0.0 mm/h
Wednesday, 27 February 2013			
00h	17.7 kts	1033.3 mb	0.0 mm/h
06h	19.9 kts	1032.2 mb	0.1 mm/h
12h	19.0 kts	1032.0 mb	0.1 mm/h
18h	19.3 kts	1030.0 mb	0.0 mm/h
Thursday, 28 February 2013			
00h	20.2 kts	1029.5 mb	0.0 mm/h
06h	19.9 kts	1027.2 mb	0.1 mm/h

Fig 7.5 *Forecasts at a point on an app (using PocketGrib)*

Processed or selected GRIB data

There are several websites showing charts of winds, isobars, sea state, etc for pre-selected areas and times. These may or may not be useful to your particular sailing needs. They may contain exactly the same information as in the basic GRIB files shown above; alternatively, they may contain selections of that data or be processed in some form or other. The advantage of using a browser is ease of access. The downside is that the information cannot easily be saved for reference when offline, which reduces its usefulness. The display in Fig 7.6 has winds at the grid points provided by NOAA.

Some websites interpolate the data to specific locations. There is nothing wrong with this, but do not be deceived into thinking that the forecasts are more site-specific or more accurate. Such data can only be derived by interpolation from the 0.5° grid (~ 50km).

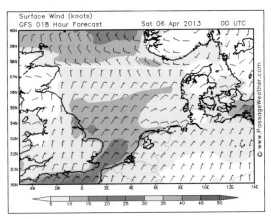

Fig 7.6 *A pre-selected area on a browser*

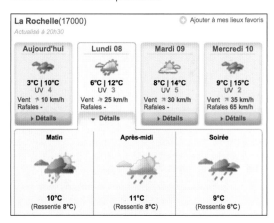

Fig 7.7 *Points forecast on a browser (using WindGuru)*

In the same category, there are forecasts by national weather services, partly in graphical form and part alphanumeric, from their operational NWP models. The display in Fig 7.8 is from the Météo France automated forecast for La Rochelle. This is an interpolation from their operational model with a 10km grid. Other countries have similar products.

Fig 7.8 *Localised Met service forecasts*

Fine-scale processed GRIB data

Output from some regional scale NWP models can be obtained online at no charge. Some sources are given in the Appendix. Usually these are only available on a browser, but some may be obtained using a tablet app or as email attachments.

Fig 7.9 is a display on the Danish Met Service site.

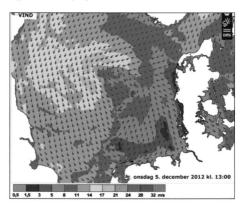

Fig 7.9 *Mesoscale forecast from Denmark*

The UK Met Office has a free app for tablets that gives short period forecasts just for the United Kingdom. Wind and temperature data is given at specific locations rather than on a grid. Rainfall and cloud cover are shown in a graphic display. The same presentations are also on their website (see Fig 7.10).

A detailed prediction of rain areas looks like Fig 7.11.

The definition of rain areas will be reasonably good but the detail will become less accurate over the forecast period. Shower locations will not be good after the first few hours.

Fig 7.10 *Short period detailed forecast of wind*

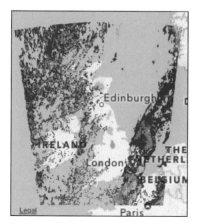

Fig 7.11 *Short period detailed forecast of rain*

On prepayment

Several commercial firms run mesoscale models. Some are listed in the Appendix. These start with output from a global model run by a major national weather service (usually the USA). Firms sometimes have trial offers so that potential users can ascertain which, if any, provides better guidance than free services.

Consultants

Weather consultants will probably have more information than the sailor, more time to digest it and, in most cases, more experience and skill. Whether or not you use a consultant will depend upon your own experience, confidence and ability. Some experienced ocean sailors prefer to use GRIBs, perhaps with routeing software and synoptic charts. Others prefer to take advice from a consultant.

> Whatever you do, bear in mind that safety of vessel and crew are the sole responsibility of the skipper. Any forecasts, from whichever source, should only be regarded as advisory. If you use a consultant, choose one who uses human judgement together with NWP output, rather than providing a purely objective service.

Actual weather reports

The most convenient source for latest actual weather reports is the internet, although practice varies greatly from country to country. The UK, Irish and German weather services have pages of actual reports updated frequently. NOAA provides reports from ships on passage and from data buoys for all parts of the world. Some coastguard broadcasts include latest actual reports.

A complete source of actual reports from locations over Europe and eastern parts of the USA is available on the XCWeather site.

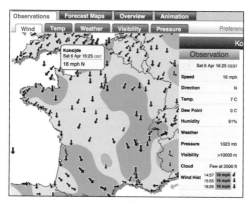

Fig 7.12 *XCWeather actual report*

Some radio-teletype broadcasts include weather actuals using meteorological five-figure code forms. There are software packages to translate the data. Some packages will analyse the data and produce charts of isobars. Use these with care as the data is not always quality-controlled and the commercial packages can all too easily produce nonsense. When using data themselves, weather services put much effort into ensuring data quality.

Summary

There are, essentially, two kinds of forecast:

◆ GMDSS forecasts, usually in text, sometimes in chart form. These are general in nature but contain human input. For safety reasons, a skipper should ensure that they are always available.

◆ GRIB files and products, usually available in some charted form but sometimes for specific locations. These provide much information that supplements and complements GMDSS forecasts.

The prudent sailor will use both.

Fig 7.13 *VHF (inshore use) and NAVTEX (when over 30 miles offshore) provide vital safety information*

Radio or internet?

There are two broad, main types for the reception of marine safety information including weather forecasts and warnings. First, there are terrestrial radio or satellite broadcast systems. Secondly, there is the internet, using terrestrial telephone or data networks. A third category for those going offshore is the use of internet access via HF radio. Most of you reading this book will not need to use the third method.

GMDSS information is broadcast over VHF, MF and HF radio including NAVTEX, an MF radio text system, and Inmarsat, a satellite text transmission system.

Marine VHF and MF

VHF radio has a line-of-sight range determined mainly by the height of the transmitting aerial. Broadcasts are usually preceded by a call, usually on Channel 16, using the highest aerial available at the transmitter. Broadcasts on working channels use aerials mounted lower down in order to restrict range and avoid interference. That is why you may hear a distant station announcement on Ch 16 but not the following broadcast. Information, mainly coastal waters forecasts and warnings, will relate mainly to the area of VHF coverage.

The golden rule is to have the radio switched on the calling channel at all times to ensure that you hear all routine and warnings broadcasts. If you know the working channel, you can use the dual-watch facility. Never leave the radio switched only to a working channel.

Bear in mind that:

◆ Reception can be poor because of hills, other obstructions and distance from the transmitter.

NAVTEX

- ◆ You can be distracted or just too busy, perhaps avoiding some hazard.
- ◆ Unless you write the information down as it is received, it can easily be forgotten.
- ◆ To use a marine VHF radio transmitter requires certification, the Short Range Certificate, by a relevant national authority. For leisure sailors in the UK, that is the RYA.

Outside VHF range, up to around 200 miles offshore, reception of voice broadcasts requires use of MF/SSB radio. No certification is needed for a receive-only set but relatively few leisure vessels carry or use such equipment.

NAVTEX

With a few exceptions, notably Australia, New Zealand and Arctic waters, GMDSS offshore weather information and navigation warnings are broadcast on NAVTEX. This telex-over-radio system uses MF/SSB radio transmissions on a common frequency, worldwide, of 518 kHz. It is intended for use at sea from the fairway buoy to at least 200 miles out. Reception may be possible in harbour or near the coast but do not rely on this. There will be other sources of information.

Broadcasts are always available in English and relate to the reception area. Weather information is for offshore (shipping) forecast areas. A second frequency of 490 kHz is used to meet national needs; the UK, Ireland and Germany use this for inshore forecasts. Several other countries use it for national language versions of 518 kHz broadcasts.

The main advantage of NAVTEX is its automatic reception using fairly low-cost equipment. The choices of frequency and the low transmitting power were intended to limit the range and minimise interference from more distant stations broadcasting at the same time.

Although the technology dates back many years, the system is widely used and unlikely to be replaced in the near future. Transmissions are very slow. Under international law, NAVTEX is obligatory for all training vessels going over 30 miles offshore. The UK marine authorities recommend its use.

As the set will receive all signals on 518 kHz within range, the operator simply selects the desired stations and message types. Both have single letter identifiers. Message types of most interest to the leisure sailor are:

A Navigation warnings
B Weather warnings
C Ice information
D Emergency messages
E Weather forecasts
L Additional navigation warnings

These message types should not be rejected.

A weather forecast (identifier letter E) from the British transmitter Cullercoats (identifier letter G) would start GENN where NN is the message number from 01 to 99.

Within each METAREA, broadcast schedules are in ten-minute time slots according to the station identifier letter. The sequence is then repeated throughout the 24 hours. Sources of information on stations and their indicator letters are in the Appendix.

Coordination under the International Maritime Organization (IMO) ensures that two stations with the same identifier letters – for example, Svalbard, Norway, and Corsen, France, both letter A – are far enough apart to minimise risks of inter-station interference. They also try (but do not always succeed) to ensure that stations broadcasting in succession are sufficiently far apart for there to be no interference should one station overrun its allocated time.

The main reception problem with NAVTEX can arise because of changes in radio propagation at night. Then, as the sun sinks below the horizon, reduction in ionisation can lead to anomalous propagation from distant stations, perhaps well over a thousand miles away. This is shown diagrammatically here:

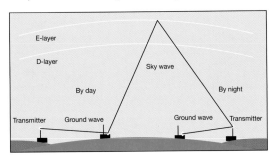

Fig 8.1 *NAVTEX anomalous propagation*

One example is Cullercoats in Northumberland and Tarifa, southern Spain, both letter G. Another is Valentia, south-west Ireland, and Toulon, Mediterranean France, both letter W.

MF/HF/SSB radio
Some coastal sailors receive radio-facsimile and radio-teletype broadcasts using a laptop PC to decode the

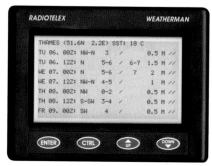

Fig 8.2 *The NASA Weatherman*

signals. Synoptic charts can be received from weather services worldwide. Some text forecasts are also available, notable from the USCG for oceans, and from Germany for the North, Baltic and Mediterranean Seas and the English Channel. The NASA Weatherman receives the European information automatically.

The internet

Although not always available, this provides a vast amount of information, although much of it is the same material presented in different formats. The major benefits of using the internet are:

◆ Being able to obtain information at convenient times.

◆ The amount of information available.

◆ Being able to read rather than listen to information when it is in a foreign language.

◆ Storing information for later reference and use when offline.

The internet

There are some problems with internet forecasts:

◆ Reception, although often good, is not always possible or reliable. Even in developed countries, coverage will not be universal.

◆ Reception out at sea will be either limited with a terrestrial-based system or (currently) costly with a satellite one.

◆ The alternative of using HF/SSB radio and a modem requires the Long Range Certificate.

◆ The internet is not an operational service in the GMDSS sense. Outages are always possible, although rare, but URLs can change without notice, even those used by national weather services.

◆ Importantly, there is currently no way by which the sailor can be made aware, automatically, of warnings.

Internet access – coastal and ashore

For coastal sailors, there are several options:

◆ Internet cafés.

◆ WiFi in harbours, or, with the password and a signal booster, from a known hotspot. Cost varies greatly from being free to several euros an hour.

◆ Cellular data networks ashore and up to 5 or so miles offshore, but more with a dedicated aerial. Costs vary from country to country and between suppliers.

These data services, using a laptop or a tablet computer, will meet most needs.

With no WiFi or cellular data network available, a General Packet Radio Service (GPRS) enabled cellphone or one just using the Global System for Mobile Communications (GSM) would allow the sending and receiving of brief emails, preferably in plain text format. To avoid large cellphone costs, take care to block large emails, say > 50 KB; also, with GPRS, to disable all automatic updates. Those important issues are beyond the scope of this book.

Internet access – long range

For those not using HF/SSB radio, the alternative involves satellite network – Iridium, Inmarsat, Thuraya or Globalstar. Broadband systems are now being developed and will be cost-effective solutions for some.

The choice will depend on how much long-distance sailing will be done, technical ability and consideration of costs. Those planning to make ocean passages are advised to take time to consider the options, then install, test and become familiar with the equipment. There are many aspects to take into account, which are beyond the scope of this book. The MailASail.com website is a good starting point. Advice from a supplier who deals with a range of such systems is sensible. They will provide advice depending upon individual needs.

The technology is developing rapidly; costs of equipment and communications will decrease in time. As a result, it is likely that satellite systems will eventually supersede terrestrial telephone and data networks.

Weather lore and rules of thumb

Weather folklore is based on a mass of experience over thousands of years and there must be some real information there. 'Red sky at night' is one of the more common and was known to the ancient Greeks. It does have some scientific justification but is not always a good guide.

Admiral Fitzroy, first head of the UK Met Office, drew up a list of rules based on watching the wind, the sky and the barometer. Many of these were the basis of sailors' rhymes such as: 'Wind before rain, let your topsail fill again. Rain before wind, sheets and topsails mind.' Another popular aphorism is, 'Mackerel skies and mare's tails make tall ships carry small sails'.

However, although there is often some foundation for such 'rules', they are of limited value to sailors because so much more and better information is now available. A particular shortcoming is that they depend on a snapshot at one place and without knowledge of what is happening over a wider area.

Fig 9.1 *Red sky at night…* © Maggie White

Even though the sight of cirrus spreading from the north-west is a good indicator of an approaching warm front, this is not a great help. It does not provide information on when the front will arrive, how strong the wind will be or how far it will back, and so on. For much of the time, the most a single observation can do is provide some confirmation of what should already be known. At best, it should prompt the sailor to look at the forecast more carefully, should what is seen be a surprise.

Nowcasting

This is a term used by national weather services to describe very short period forecasts based, to a large extent, on current data from weather radar and satellite images coupled with data from observing sites and fine-scale NWO models. Similarly, sailors can make some useful short-term forecasts using their own observations.

Forecasts should have given some indication of strong winds days ahead, but a rapidly rising or falling barometer should provide a few hours' notice that strong winds are about to occur.

Rough guides of expected wind speed are for changes in pressure of:

◆ 6 hPa in three hours – gale force 8.

◆ 10 hPa in three hours – storm force 10.

Of more value is watching the weather for immediate problems. A dinghy sailor will see a gust approaching and be prepared to free the mainsail as it hits or bear off a little, if possible. The cruising sailor can do the same, but watching for bigger squalls. A patch of more disturbed sea upwind or that line of roll cloud must mean something.

Watch convective cloud for the first signs that the tops are starting to freeze as an indicator of showers or thunderstorms. Even just large cumulus clouds mean that there is vigorous convection and gusts are likely. When visibility is poor, watching the horizon and any other vessels can show just how bad the visibility is becoming. That can be deceptive, so switch on the radar and see how far away you can really see a coast or a ship.

Although what is seen should have been mentioned or been apparent from the forecast, weather is such that a surprise might still be in store. Occasionally, something will happen to catch even the most experienced sailors or meteorologists off-guard. Watching the sky and trying to relate what is seen to what is being experienced is always interesting if you have an enquiring mind. The following photographs may be of clouds that are rarely encountered but they provide some ideas of what to look for and how much information might be there.

Examples of unusual cloud patterns

There is always a reason for everything that we see happening in the atmosphere. But at first or even second sight, those reasons might not be obvious. These are some examples – and explanations – of some of the more atypical patterns that you might encounter.

Mountain waves

At first sight, these clouds may seem to be of academic interest to the sailor. They are formed by standing waves and can help glider pilots to soar. However, the effects at sea level in the wave troughs might be some unexpectedly strong winds.

Examples of unusual cloud patterns

Fig 9.2 *Wave clouds* © *Vyv Cox*

Fig 9.3 *More wave clouds* © *Vyv Cox*

Examples of unusual cloud patterns

Fig 9.4 *Bora cloud*

Bora cloud

The bora is the name given to the strong northeast wind that occurs in the Adriatic. This very distinctive line of cloud can be seen when there is a bora. They are a good warning to stay in port.

Mammatus cloud

These pendulous-looking clouds indicate that heavy rain has been falling in a shower. The rain might have ceased but they are a warning that there are heavy showers around. Sudden strong winds may occur in the downdraughts.

There is a reason for everything you see in the sky and experience on the water, although it may be difficult to understand and often of little practical value. From time to time, however, careful observation and a little thought can provide a useful lesson – maybe by the hard way.

Examples of unusual cloud patterns

Fig 9.5 *Mammatus cloud © Robin Atkinson*

So, keep watching; keep learning; do not be afraid to ask others who might have had similar experiences or who might offer explanations and advice. Like sailing in general, there will always be something new that you have not encountered before.

Fig 9.6 *Mackerel skies © Maggie White*

Acronyms and abbreviations

You will find most of these in this book, and there are some additional ones that you might encounter.

AEMET	Spanish weather service
AIS	Automatic [ship] Identification System
ALADIN	A French acronym similar to HIRLAM (see below)
CAPE	Convective Available Potential Energy (a lightning risk index)
COAMPS	[US Navy] Coupled Ocean/Atmosphere Mesoscale Prediction System
DWD	German weather service
ECMWF	European Centre for Medium-Range Weather Forecasts
GMDSS	Global Maritime Distress and Safety System
GFS	Global Forecast System (NOAA NWP model)
GPRS	General Packet Radio Service (for internet access over a cellphone)
GRIB	Gridded Binary (compressed data files used by NMS to exchange NWP output)
GSM	Global System for Mobile communications (cellphone)
HF, MF, VHF	High, Medium, Very High Frequency (radio transmissions)

HIRLAM	High Resolution Limited Area Model (NWP)
IMO	International Maritime Organization
HMCG	UK Coastguard
JCOMM	Joint Technical Commission for Oceanography and Marine Meteorology
KNMI	Dutch weather service
LRC	Long Range [radio] Certificate (for MF/HF/SSB and satellite)
MCA	[UK] Maritime and Coastguard Agency (parent of HMCG)
Met Éireann	Irish weather service
METAREAs	Geographical sea regions for the purpose of coordinating the transmission of meteorological information to offshore sailors

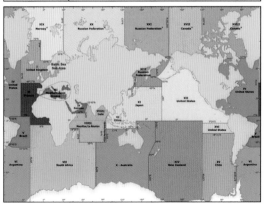

Fig 10.1 *A map of all METAREAs*

Acronyms and abbreviations

MRCC	Maritime Rescue Coordination Centre
NAM	North American Mesoscale [Forecast System]
NAVAREAs	See **METAREAs**
NAVTEX	Navigational Telex
NMS	National Meteorological Service
NOAA	[US] National Oceanic and Atmospheric Administration
NWP	Numerical Weather Prediction
NWS	[US] National Weather Service
Poseidon	Greek marine research group mesoscale model
RYA	Royal Yachting Association
SOLAS	[International Convention for] Safety of Life at Sea (parent of the GMDSS)
SRC	Short Range [radio] Certificate (for VHF)
SSB	Single Side Band (radio transmission)
UM	Unified Model (UK Met Office NWP model used in mesoscale and global form)
WRF	Weather Research and Forecasting (mesoscale model)

Meteorological terms

Adiabatic cooling/ warming	Ascending/descending air expands/ contracts leading to cooling/ warming although there has been no external cooling/heating.
Buys Ballot's law	In the northern hemisphere, with your back to the wind, pressure is lower to your left than to your right.
Condensation level	The height at which rising air is cooled to its dewpoint temperature.
Convective clouds	These form when warming of the surface cause bubbles of air to rise.
Coriolis effect	The effect of the spin of the earth on the wind: zero at the equator, maximum at the poles (see Buys Ballot's law).
Cumulus	Heaped cloud formed by convection.
Cumulonimbus	Large cumulus cloud giving heavy, often thundery showers.
Cirrus	High wispy cloud.
Dewpoint	The temperature at which condensation occurs when the air is cooled.
Drizzle	Light, steady precipitation, normally from fairly shallow, low cloud found in warm sectors.

Meteorological terms

Friction layer	The lowest levels of the atmosphere where the wind is affected by surface friction. Typically about 1,000m deep.
Ensemble	A set of NWP forecasts produced with small random variations in the original analysis.
Geostrophic wind	The wind that occurs with straight isobars and above the friction layer.
Gradient wind	The wind that occurs above the friction layer allowing for curvature of the isobars.
hPa	Hectopascal – unit of pressure (1 hPa = 1 millibar)
Jet stream	High-level strong winds often associated with fronts.
Latent heat	The heat required to melt ice or evaporate water and released during condensation and freezing.
Mesoscale	Weather on scales below that seen on synoptic charts. Sea breezes, showers and thunderstorms are examples. Applied to small-scale NWP.
Rain	Used by itself to mean precipitation usually from layered – i.e. not convective – cloud.
Showers	Precipitation from convective cloud.

Stable / stability	The air is stable if convection cannot occur or is limited in vertical extent; typically in a warm sector or around a high pressure area.
Stratosphere	The atmosphere above the tropopause.
Stratus	A layer of low cloud – sometimes nimbostratus. Also cirrostratus and altostratus – layers of high- and medium-level cloud.
Synoptic scale	Large-scale weather that can be analysed on weather charts covering large areas.
Tropopause	The top of the atmosphere as we see it. About 8 or 9km high in polar regions and 16 or 17km near the equator. Usually marked by a temperature inversion and the maximum height that convection can reach.
Troposphere	The atmosphere below the tropopause.
Unstable / instability	The air is said to be unstable when convection from the ground leads to cumulus cloud forming; typically in air from the poles.

Terms used in marine weather forecasts

Beaufort wind scale

With minor variations in the descriptive terms, the wind scale is agreed internationally.

Force	Knots	Description
0	<1	Calm
1	1– 3	Light air
2	4– 6	Light breeze
3	7–10	Gentle breeze
4	11–16	Moderate breeze
5	17–21	Fresh breeze
6	22–27	Strong breeze
7	28–33	Near gale
8	34–40	Gale
9	41–47	Severe gale
10	48–55	Storm
11	56–63	Violent storm
12	64+	Hurricane force

Douglas sea state

This table relates expected wave height to Beaufort forces. The wave heights shown here are those likely in the absence of any swell. The Douglas sea state scale is not used in UK Met Office forecasts but is used by some other countries.

Force	Mean wave (m)	Douglas	Sea state
0	–	0	Calm
1	0.1	1	Calm
2	0.2	2	Smooth
3	0.6	3	Slight
4	1.0	3–4	Slight–moderate
5	2.0	4	Moderate
6	3.0	5	Rough
7	4.0	5–6	Rough–very rough
8	5.5	6–7	Very rough–high
9	7.0	7	High
10	9.0	8	Very high
11	11.5	8	Very high
12	14+	9	Phenomenal

Terms used in marine weather forecasts

Gale and strong wind warnings

Depending on the type of forecast, warnings can be issued for force 6 and above. In UK shipping forecasts, a gale warning includes gusts to 41 knots even when the mean speed is less than force 8.

For force 12, the term used is 'hurricane force 12'; the term 'hurricane' on its own means a tropical cyclone over the Atlantic and eastern Pacific oceans. These do not occur in European waters.

Wind direction

Direction is given as that from which the wind is coming. Note that current is expressed as the direction of the current. The following terms are used:

Veering	Direction changing clockwise, e.g. SW to W or E to SE
Backing	Direction changing anticlockwise, e.g. W to SW or SE to E
Becoming cyclonic	Direction changing across the path of a low within the forecast area

Visibility

Fog	Less than 1,000m
Poor	1,000m to 2nm
Moderate	1nm to 5nm
Good	More than 5nm

Terms used in UK Met Office forecasts

Gale warning timings

These terms are defined for gale warnings but also appear in marine forecasts.

Imminent	Within 6 hours of time of issue
Soon	Within 6–12 hours of time of issue
Later	More than 12 hours from time of issue

NOTE: The UK Met Office shipping forecast may sometimes use the words 'perhaps gale 8 later', but not issue a gale warning. That is when there is some doubt about the wind strength. For periods up to 12 hours ahead, any mention of gale in the forecast must be accompanied by a gale warning.

Movement of pressure systems

Slowly	Less than 15 knots
Steadily	15–25 knots
Rather quickly	25–35 knots
Rapidly	35–45 knots
Very rapidly	More than 45 knots

Pressure tendency in station reports

Values relate to the past three hours and qualify the description of falling or rising.

Description	Falling or rising
More slowly	Progressively slower rate through the preceding 3 hours
Slowly	0.1–1.5 hPa
N/A	1.6–3.5 hPa
Quickly	3.6–6.0 hPa
Rapidly	More than 6.0 hPa
Now falling (rising)	A change from rising (falling) to falling (rising)

Sources of information

Much information is online and this list is indicative rather than exhaustive. URLs are not given because they tend to change; to find links, use a search engine. There are many acronyms, abbreviations and 'trade names' in the following pages. These will be recognised by search engines.

Radio schedules and forecast areas

Many national weather services and marine authorities issue schedules of weather broadcasts but these are not always easy to find. This list will help as a start:

◆ Admiralty List of Radio Signals Vol.3 Part 1. Comprehensive and worldwide. Hard copy only.

◆ DWD. Radio facsimile and RTTY for Europe and the Mediterranean. Online.

◆ ICS Electronics Ltd. Database of NAVTEX stations worldwide. Online.

◆ JCOMM for METAREAs NAVTEX sea areas. Online.

◆ Malin Head Coastguard website.

◆ MCA Maritime Safety Information. Online and paper.

◆ Météo-France. Online and paper.

◆ NOAA NWS, VHF channels and MF frequencies.

◆ NOAA NWS. Radio facsimile stations worldwide. Online.

◆ Reeds and other almanacs. Hard copy, desktop version for PC and Mac, and iPad App.

◆ RYA booklet G5. Europe & Mediterranean. Hard copy only.

◆ William Hepburn, NAVTEX broadcasts worldwide.

Sources of information

Texts of broadcast and other forecasts

There are many sources of URLs for GMDSS forecasts. Start here:

◆ JCOMM. High seas and some offshore forecast texts online.

◆ Many national weather service websites.

◆ Marina and yacht club notice boards.

Weather charts online

Some NMS provide synoptic charts. These are sources of some of the easiest to obtain:

◆ ECMWF. Global coverage – 10-day forecasts.

◆ KNMI. Eastern N. Atlantic, Europe/Mediterranean – 3-day forecasts.

◆ NOAA NWS. All northern hemisphere, including wave charts.

◆ UK Met Office. N. Atlantic/Europe – 5-day forecasts.

Sources of free GRIB output and products

Except where noted, these refer to the NOAA global model, the GFS.

Direct download to a PC

◆ zyGrib. Many parameters. Good for CAPE and wave data.

◆ Ugrib. Wind, pressure rainfall only.

◆ NOAA and Environment Canada. Only for the more computer-literate sailors.

Tablet apps

◆ iGrib. iPad. Several parameters.

◆ MobileGrib, Android. Several parameters, including waves and CAPE.

◆ PocketGrib, iPad and Android. Several parameters, including CAPE and wave models. GFS and COAMPS.

◆ SailGrib, Android. Most parameters.

◆ Weather4D, iPad. Several parameters, including waves and CAPE.

◆ Weathertrack, iPad. Several parameters. Good CAPE display. Wave models. GFS and COAMPS.

More apps will be developed as the number of tablets increases.

Email services

◆ Saildocs.com. Several parameters, including wave data.

◆ MailASail.com. Fairly basic, easier to use than Saildocs. Wave data.

◆ Weather4D app has a facility to get GRIBs via email from Saildocs.

Sources of information

Browser services

- Magicseaweed. Wind and sea state.

- PassageWeather. Wind, pressure, sea state.
  Also as an app.

- WeatherOnline. Several models, many parameters.

- Weatherweb. Several models.

- Windguru. Interpolated to specific locations.

- XCWeather. Fairly limited but excellent for getting latest actual weather.

Regional GRIB models

Free

Most of these are for Europe and the Mediterranean.

By transfer direct to a PC

- Norwegian Met Service. HIRLAM 2-day forecasts for pre-selected European and Scandinavian areas.

Email

- Saildocs for COAMPS.

By browser

- Athens University. E. Mediterranean.

- Croatian Met Service. Adriatic.

- Danish Met Service (DMS). North Sea and S. Baltic.

- Greek Met Service (HNMS). Greek waters.

- Icelandic Met Service. Icelandic waters.

- LAMMA. Seas to west of Italy.

- ◆ Norwegian Met Service.
- ◆ PassageWeather (COAMPS for Europe, NAM for USA).
- ◆ Poseidon. E. Mediterranean.
- ◆ Spanish Met Service (AEMET). Spanish waters, SE N. Atlantic and W. Mediterranean.

GRIB services on prepayment

Most of these provide GFS and mesoscale output.

- ◆ Clearpointweather.com
- ◆ PredictWind.com
- ◆ Theyr.com
- ◆ Theyr.tv
- ◆ Weather365.net